Degree Theory for Operators of Monotone Type and Nonlinear Elliptic Equations with Inequality Constraints

of the
American Mathematical Society

Number 915

Degree Theory for Operators of Monotone Type and Nonlinear Elliptic Equations with Inequality Constraints

Sergiu Aizicovici
Nikolaos S. Papageorgiou
Vasile Staicu

November 2008 • Volume 196 • Number 915 (second of 5 numbers) • ISSN 0065-9266

American Mathematical Society
Providence, Rhode Island

2000 *Mathematics Subject Classification.* Primary 35J85, 35J60.

Library of Congress Cataloging-in-Publication Data

Aizicovici, Sergiu, 1948–
  Degree theory for operators of monotone type and nonlinear elliptic equations with inequality constraints / Sergiu Aizicovici, Nikolaos S. Papageorgiou, and Vasile Staicu.
    p. cm. — (Memoirs of the American Mathematical Society, ISSN 0065-9266 ; no. 915)
  "November 2008, volume 190, number 915 (second of 5 numbers)."
  Includes bibliographical references.
  ISBN 978-0-8218-4192-1 (alk. paper)
  1. Differential equations, Elliptic. 2. Differential equations, Nonlinear. 3. Variational inequalities (Mathematics) 4. Topological degree. 5. Monotone operators. I. Papageorgiou, Nikolaos Socrates. II. Staicu, Vasile. III. Title.

QA377.A529 2008
515′.3533—dc22   2008030040

Memoirs of the American Mathematical Society

This journal is devoted entirely to research in pure and applied mathematics.

Subscription information. The 2008 subscription begins with volume 191 and consists of six mailings, each containing one or more numbers. Subscription prices for 2008 are US$675 list, US$540 institutional member. A late charge of 10% of the subscription price will be imposed on orders received from nonmembers after January 1 of the subscription year. Subscribers outside the United States and India must pay a postage surcharge of US$38; subscribers in India must pay a postage surcharge of US$43. Expedited delivery to destinations in North America US$53; elsewhere US$130. Each number may be ordered separately; *please specify number* when ordering an individual number. For prices and titles of recently released numbers, see the New Publications sections of the *Notices of the American Mathematical Society*.

Back number information. For back issues see the *AMS Catalog of Publications*.

Subscriptions and orders should be addressed to the American Mathematical Society, P. O. Box 845904, Boston, MA 02284-5904, USA. *All orders must be accompanied by payment.* Other correspondence should be addressed to 201 Charles Street, Providence, RI 02904-2294, USA.

Copying and reprinting. Individual readers of this publication, and nonprofit libraries acting for them, are permitted to make fair use of the material, such as to copy a chapter for use in teaching or research. Permission is granted to quote brief passages from this publication in reviews, provided the customary acknowledgment of the source is given.

Republication, systematic copying, or multiple reproduction of any material in this publication is permitted only under license from the American Mathematical Society. Requests for such permission should be addressed to the Acquisitions Department, American Mathematical Society, 201 Charles Street, Providence, Rhode Island 02904-2294, USA. Requests can also be made by e-mail to reprint-permission@ams.org.

Memoirs of the American Mathematical Society (ISSN 0065-9266) is published bimonthly (each volume consisting usually of more than one number) by the American Mathematical Society at 201 Charles Street, Providence, RI 02904-2294, USA. Periodicals postage paid at Providence, RI. Postmaster: Send address changes to Memoirs, American Mathematical Society, 201 Charles Street, Providence, RI 02904-2294, USA.

© 2008 by the American Mathematical Society. All rights reserved.
Copyright of individual articles may revert to the public domain 28 years after publication. Contact the AMS for copyright status of individual articles.
This publication is indexed in *Science Citation Index*®, *SciSearch*®, *Research Alert*®, *CompuMath Citation Index*®, *Current Contents*®/*Physical, Chemical & Earth Sciences*.
Printed in the United States of America.

∞ The paper used in this book is acid-free and falls within the guidelines established to ensure permanence and durability.
Visit the AMS home page at http://www.ams.org/

10 9 8 7 6 5 4 3 2 1   13 12 11 10 09 08

Contents

Chapter 1. Introduction   1

Chapter 2. Mathematical Background   5

Chapter 3. Degree Theoretic Results   13

Chapter 4. Variational-Hemivariational Inequalities   27

Chapter 5. Hemivariational Inequalities with an Asymmetric Subdifferential   51

Bibliography   69

Abstract

In this paper, in the first part, we examine the degree map of multivalued perturbations of nonlinear operators of monotone type and we prove that at a local minimizer of the corresponding Euler functional, this degree equals one. Then we use this result to prove multiplicity results for certain classes of unilateral problems with nonsmooth potential (variational-hemivariational inequalities). Also we prove a multiplicity result for a nonlinear elliptic equation driven by the p-Laplacian with a nonsmooth potential (hemivariational inequality) whose subdifferential exhibits an asymmetric asymptotic behavior at $+\infty$ and $-\infty$.

Received by the editor November 10, 2005.
1991 *Mathematics Subject Classification*. 35J85, 35J60.
Key words and phrases. operator of monotone type, degree map, minimizer, nonsmooth potential, unilateral problem, multiple solutions.
This paper was completed while the first two authors were visiting the University of Aveiro. The hospitality and financial support of the host institution are gratefully acknowledged. The third author acknowledges partial financial support from the Portuguese Foundation for Sciences and Technology (FCT) under the project POCI/MAT/55524/2004.

CHAPTER 1

Introduction

In 1982, Amann [**1**] established that the Leray-Schauder degree of an isolated local minimum of a C^1 functional in a Hilbert space equals one. Subsequently, the result of Amann was extended by Hirano [**21**] to smooth functionals whose gradient is an operator of type $(S)_+$ on a Hilbert space.

This extension was based on the degree map introduced by Browder [**8**], which covers large classes of operators of monotone type from a reflexive Banach space X to its dual X^*. Very recently, Kobayashi-Otani [**27**], presented a further extension of Amann's result to operators of the form $\nabla\varphi + \partial\psi$, with φ a C^1 functional on a reflexive Banach space X, whose gradient is an operator of type $(S)_+$ and $\psi \in \Gamma_0(X)$, i.e., $\psi : X \to \overline{\mathbb{R}} = \mathbb{R} \cup \{+\infty\}$ is a proper, convex and lower semicontinuous functional, which is not identically $+\infty$.

The work of Kobayashi-Otani paved the way for a degree theoretic treatment of nonlinear variational inequalities, based on the degree map of Browder. In this paper, we extend the work of Kobayashi-Otani [**27**] and we prove a result concerning the degree of an isolated minimizer for Euler functionals of the form $\varphi + \psi$, where φ is nonsmooth and locally Lipschitz and $\psi \in \Gamma_0(X)$. In this case the operator under consideration is $\partial\varphi + \partial\psi$, with $\partial\varphi$ the generalized subdifferential of φ and $\partial\psi$ the convex subdifferential of ψ. The calculation is based on an extension of Browder's degree map due to Hu-Papageorgiou [**24**]. This generalization permits the study of nonlinear variational inequalities with a nonsmooth potential function (variational-hemivariational inequalities).

In variational-hemivariational inequalities, we have the combined effects of two types of unilateral constraints. The first comes from the maximal monotone term which is not in general everywhere defined (variational inequality) and the second from the nonmonotone but everywhere defined term $\partial\varphi(x)$ (hemivariational inequality).

Hemivariational inequalities are a new type of variational expressions which arise naturally in problems of Mechanics and Engineering, if one wants to consider more realistic laws which are multivalued (nonsmooth potential) and nonmonotone (nonconvex potential). Concrete applications of hemivariational inequalities can be found in the book of Naniewicz-Panagiotopoulos [**41**].

The degree theoretic approach was first used in semilinear variational inequalities by Szulkin [**43**], [**44**], Miersemann [**37**] and Kucera [**28**]. More recently, important contributions in the direction of nonlinear problems, were made by Le [**32**], [**33**].

In the next Chapter, for the convenience of the reader, we recall some basic definitions and facts from nonsmooth and nonlinear analysis, which we will need in the sequel. Our basic references are the books of Barbu [**5**], Gasinski-Papageorgiou [**18**], Denkowski-Migorski-Papageorgiou [**12**], [**13**] and Zeidler [**46**].

In Chapter 3 we prove the extension of Amann's theorem concerning the Leray-Schauder degree of a C^1- functional defined on a Hilbert space, at an isolated local minimum. In our formulation (see Theorem 8), the result is proved for operators which are sums of a generalized subdifferential and of a convex subdifferential and are defined on a reflexive Banach space which is embedded compactly and densely in $L^p(Z)$ $(1 < p < \infty)$, where Z is a bounded open subset of $\mathbb{R}^N$ with a smooth boundary.

In Chapter 4, we use the degree theoretic results of Chapter 3, to prove multiplicity results for certain nonlinear obstacle problems with nonsmooth potential (variational-hemivariational inequalities). The results of this Chapter extend the work of Kobayashi-Otani [27].

In Chapter 5, we employ degree theoretic methods to prove a multiplicity result for nonlinear hemivariational inequalities in which the subdifferential nonlinearity exhibits an asymmetric behavior as we approach $+\infty$ and $-\infty$. In fact as we move from $-\infty$ to $+\infty$ the *slopes* $\left\{\frac{u}{|x|^{p-2}x} : u \in \partial j(z,x)\right\}$ ($j(z,.)$ is the nonsmooth locally Lipschitz potential function and $\partial j(z,.)$ its generalized subdifferential, see Chapter 2) cross the principal eigenvalue $\lambda_1 > 0$ of $\left(-\triangle_p, W_0^{1,p}(Z)\right)$, where $\left(-\triangle_p, W_0^{1,p}(Z)\right)$ stands for the negative p-Laplacian on Z with zero boundary condition.

In the past, the problem of existence of multiple solutions for elliptic equations was primarily investigated in the context of semilinear problems (i.e., $p = 2$) with a smooth potential. In this direction we mention the works of Gonçalves-Miyagaki [20], Hirano [22], Landesman-Robinson-Rumbos [31], and Mizoguchi [38]. In all these works the right hand side nonlinearity exhibits a symmetric behavior asymptotically at $+\infty$ and $-\infty$. In Hirano-Nishimura [23], the authors deal with asymmetric nonlinearities using the Fucik spectrum of $\left(-\triangle, H_0^1(Z)\right)$.

Multiplicity results for problems driven by the p-Laplacian were obtained by Dinca-Jebelean-Mawhin [14], Liu-Su [36] and Li-Zhou [35]. In all these results, the right-hand side nonlinearity exhibits a symmetric behavior at $+\infty$ to $-\infty$. Dinca-Jebelean-Mawhin [14] employ the well-known Ambrosetti-Rabinowitz condition, Liu-Su [36] use Morse theory to deal with resonant problems and Li-Zhou [35] assume a symmetric (namely odd) nonlinearity $f(z,x)$ which is continuous on $\overline{Z} \times \mathbb{R}$ and use minimax methods to prove a multiplicity result.

Recently Motreanu-Papageorgiou [40] studied nonlinear elliptic equations driven by the p-Laplacian and with a nonsmooth potential function (hemivariational inequalities). They proved a multiplicity result under the condition of symmetric behavior of the subdifferential as $x \to \pm\infty$. Under their conditions, the Euler functional of the problem is coercive, which is not true in our setting. The approach of Motreanu-Papageorgiou [40] uses variational arguments and so it is completely different from ours.

Finally, we should point out that the hypotheses on the nonsmooth potential $j(z,x)$ in Chapter 5, incorporate in our framework of analysis the so called asymptotically p-linear problems (if $p = 2$, the asymptotically linear problem). This kind of problems have attracted great interest since the pioneering work of Amann-Zehnder [2]. We mention the works of Bartsch-Li [6], Dancer-Zhang [11], Stuart-Zhou [42] (for semilinear problems, i.e., $p = 2$), and of Huang-Zhou [26],

Li-Zhou [**34**], [**35**] (for nonlinear problems with the p-Laplacian). In all these problems the right-hand side nonlinearity $f(z,x)$ is in $C\left(\overline{Z}\times\mathbb{R}\right)$ and the asymptotic condition for the slopes $\frac{f(z,x)}{|x|^{p-2}x}$ requires that the limits of these slopes as x tends to 0 and to $\pm\infty$ exist and certain monotonicities and sign conditions are also imposed.

CHAPTER 2

Mathematical Background

Let X be a reflexive Banach space with norm $\|.\|$. As usual, $2^X \setminus \{\varnothing\}$ stands for the family of all nonempty subsets of X and $\overline{\Omega}$ denotes the closure of a set $\Omega \in 2^X \setminus \{\varnothing\}$. By X^* we denote the topological dual of X with norm $\|.\|_*$, and by $\langle .,. \rangle$ the duality brackets for the pair (X^*, X). Let $w(X, X^*)$ be the weak topology on X, which is known to be a Hausdorff topology. The space X endowed with the $w(X, X^*)$ topology will be denoted by X_w. The closure of a set $\Omega \in 2^X \setminus \{\varnothing\}$ with respect to the weak topology will be denoted by $\overline{\Omega}^w$. The strong (respectively weak) convergence in X or X^* will be denoted by $\to$ (respectively, $\xrightarrow{w}$). Alternatively, the symbols "s-" and respectively "w-" will also be used throughout this paper. If $x \in X$ and $C \subseteq X$ then

$$d(x, C) := \inf \{\|x - y\| : y \in C\}$$

denotes the distance from x to C.

A function $\varphi : X \to \overline{\mathbb{R}} = \mathbb{R} \cup \{+\infty\}$ is said to be proper if it is not identically $+\infty$. The effective domain of φ is the set

$$dom\ \varphi := \{x \in X : \varphi(x) < +\infty\}.$$

By $\Gamma_0(X)$ we denote the cone of all functions $\varphi : X \to \overline{\mathbb{R}} = \mathbb{R} \cup \{+\infty\}$ which are proper, convex and lower semicontinuous. If $\varphi \in \Gamma_0(X)$, then φ is locally Lipschitz in the interior of $dom\ \varphi$. In particular, it follows that if φ is convex, lower semicontinuous and everywhere finite on X (i.e., $\mathbb{R}$-valued), then φ is continuous on X. If the space X is finite dimensional, then every convex function is locally Lipschitz on the interior of its effective domain.

If C is a closed convex subset of X, the function $i_C : X \to \overline{\mathbb{R}} = \mathbb{R} \cup \{+\infty\}$ defined by

$$i_C(x) = \begin{cases} 0 & \text{if } x \in C \\ +\infty & \text{if } x \notin C \end{cases}$$

is called the *indicator function of* C and clearly if $C \neq \varnothing$, then $i_C \in \Gamma_0(X)$.

Given $\varphi \in \Gamma_0(X)$, the multifunction $\partial \varphi : X \to 2^{X^*}$ defined by

$$\partial \varphi(x) := \{x^* \in X^* : \langle x^*, y - x \rangle \leq \varphi(y) - \varphi(x) \text{ for all } y \in X\}$$

is called the *convex subdifferential* of φ. The domain $D(\partial \varphi)$ of $\partial \varphi$ is the set

$$D(\partial \varphi) = \{x \in X : \partial \varphi(x) \neq \varnothing\}.$$

If $\varphi \in \Gamma_0(X)$ is Gâteaux differentiable at $x \in X$, then $\partial \varphi(x) = \{\nabla \varphi(x)\}$. If $\varphi = i_C$ with $C \subseteq X$ nonempty, closed and convex, then

$$\partial i_C(x) = \left\{ x^* \in X^* : \langle x^*, x \rangle = \sigma(x^*; C) = \sup_{y \in C} \langle x^*, y \rangle \right\}.$$

This set is a closed convex cone, called the normal cone to C at x.

If $\varphi_1, \varphi_2 \in \Gamma_0(X)$, then, in general, we have
$$\partial \varphi_1 + \partial \varphi_2 \subseteq \partial(\varphi_1 + \varphi_2)$$
and if there exists a point $x \in dom\ \varphi_2$ such that φ_1 is continuous at x, then
$$\partial \varphi_1 + \partial \varphi_2 = \partial(\varphi_1 + \varphi_2).$$

Given an operator $A: X \to 2^{X^*}$, we define the *domain* of A by
$$D(A) := \{x \in X : A(x) \neq \varnothing\}$$
and the *graph of* A by
$$GrA := \{(x, x^*) \in X \times X^* : x^* \in A(x)\}.$$

We say that A is *monotone*, if
$$\langle x^* - y^*, x - y \rangle \geq 0 \text{ for all } (x, x^*), (y, y^*) \in GrA.$$

We say that A is *strictly monotone*, if
$$\langle x^* - y^*, x - y \rangle > 0 \text{ for all } (x, x^*), (y, y^*) \in GrA, \ x \neq y.$$

The operator A is called *maximal monotone*, if it is monotone and for $y \in X$ and $y^* \in X^*$ for which
$$\langle y^* - x^*, y - x \rangle \geq 0 \text{ for all } (x, x^*) \in GrA,$$
it follows that $(y, y^*) \in GrA$. In other words, GrA is maximal with respect to inclusion among the graphs of all monotone operators.

The convex subdifferential $\partial \varphi : X \to 2^{X^*}$ of a function $\varphi \in \Gamma_0(X)$ is a maximal monotone operator.

If $\varphi(x) = \frac{1}{2} \|x\|^2$, then $\mathcal{F}(x) = \partial \varphi(x)$ is called the *duality map* of X. So
$$\mathcal{F}(x) = \left\{ x^* \in X^* : \langle x^*, x \rangle = \|x\|^2 = \|x^*\|_*^2 \right\}.$$

If X^* is strictly convex, then the duality map is single-valued, bounded, coercive and odd.

If X and X^* are strictly convex, then the duality map is strictly monotone and bijective, with $\mathcal{F}^{-1} : X^* \to X$ being the duality map of X^*.

Finally if both X and X^* are locally uniformly convex, then the duality map $\mathcal{F}$ is a homeomorphism.

Suppose that both X and X^* are strictly convex and let $A : X \to 2^{X^*}$ be a maximal monotone operator. From the perturbation theory of maximal monotone operators (see for example Denkowski-Migorski-Papageorgiou [**12**], p.48), we know that for every $x \in X$ and every $\lambda > 0$, the operator inclusion
$$0 \in \mathcal{F}(u - x) + \lambda A(u)$$
has a unique solution $u \in D(A)$.

We define J_λ^A and A_λ by
$$J_\lambda^A(x) := u \text{ and } A_\lambda(x) := -\frac{1}{\lambda} \mathcal{F}(u - x)$$
for every $x \in X$ and every $\lambda > 0$. Both operators J_λ^A and A_λ are everywhere defined, single valued and bounded. In addition, A_λ is monotone.

An operator $S : X \to X^*$, which is single valued and everywhere defined, is said to be *demicontinuous*, if $x_n \to x$ in X, implies that $S(x_n) \xrightarrow{w} S(x)$ in X^*.

A demicontinuous, monotone operator $A : X \to X^*$ is in fact maximal monotone. Also if both X, X^* are strictly convex, then A_λ is maximal monotone.

Given $\varphi \in \Gamma_0(X)$, for every $\lambda > 0$ we define the Moreau-Yosida regularization φ_λ of φ by
$$\varphi_\lambda(x) = \inf_{y \in X}\left[\varphi(y) + \frac{1}{2\lambda}\|x-y\|^2\right].$$
Evidently, φ_λ is convex and $\mathbb{R}$-valued. Moreover, if both X, X^* are strictly convex, then φ_λ is Gâteaux differentiable and we have:

- $\varphi_\lambda(x) = \varphi(J_\lambda(x)) + \frac{1}{2\lambda}\|x - J_\lambda(x)\|^2$ for every $x \in X$ and every $\lambda > 0$,
- $\varphi(J_\lambda(x)) \leq \varphi_\lambda(x) \leq \varphi(x)$ for every $x \in X$ and every $\lambda > 0$,
- $\lim_{\lambda \to 0^+} \varphi_\lambda(x) = \varphi(x)$ for every $x \in X$,
- $(\partial \varphi)_\lambda = \partial \varphi_\lambda = \nabla \varphi_\lambda$, for every $\lambda > 0$,

where $J_\lambda := J_\lambda^{\partial \varphi}$.

An operator $A : X \to X^*$, which is single valued and everywhere defined, is said to be *of type* $(S)_+$, if for every sequence $\{x_n\}_{n \geq 1} \subseteq X$ such that $x_n \xrightarrow{w} x$ in X and
$$\limsup_{n \to \infty} \langle A(x_n), x_n - x \rangle \leq 0,$$
one has
$$x_n \to x \text{ in } X.$$
Suppose that $\{C_n\}_{n \geq 1}$ is a sequence of nonempty subsets of X. We define
$$s - \liminf_{n \to \infty} C_n = \left\{x \in X : x = s - \lim_{n \to \infty} x_n,\ x_n \in C_n,\ \forall n \geq 1\right\}$$
$$= \left\{x \in X : \lim_{n \to \infty} d(x, C_n) = 0\right\}$$
$$s - \limsup_{n \to \infty} C_n = \left\{x \in X : x = s - \lim_{k \to \infty} x_{n_k},\ x_{n_k} \in C_{n_k},\ n_k < n_{k+1},\ \forall k \geq 1\right\}$$
$$= \left\{x \in X : \liminf_{n \to \infty} d(x, C_n) = 0\right\}$$
$$w - \limsup_{n \to \infty} C_n = \left\{x \in X : x = w - \lim_{k \to \infty} x_{n_k},\ x_{n_k} \in C_{n_k},\ n_k < n_{k+1},\ \forall k \geq 1\right\}.$$

Clearly, we have
$$s - \liminf_{n \to \infty} C_n \subseteq s - \limsup_{n \to \infty} C_n \subseteq w - \limsup_{n \to \infty} C_n.$$
We say that the sequence $\{C_n\}_{n \geq 1}$ *converges to C in the Kuratowski sense*, denoted by $C_n \xrightarrow{K} C$ if and only if
$$C = s - \liminf_{n \to \infty} C_n = s - \limsup_{n \to \infty} C_n.$$
We say that the sequence $\{C_n\}_{n \geq 1}$ *converges to C in the Mosco sense*, denoted by $C_n \xrightarrow{M} C$ if and only if
$$C = s - \liminf_{n \to \infty} C_n = w - \limsup_{n \to \infty} C_n.$$

Given a function $\varphi : X \to \overline{\mathbb{R}} = \mathbb{R} \cup \{+\infty\}$, the *epigraph of* φ, denoted by epi φ, is the set
$$\text{epi } \varphi = \{(x, \lambda) \in X \times \mathbb{R} : \varphi(x) \leq \lambda\}.$$
If $\varphi_n, \varphi : X \to \overline{\mathbb{R}} = \mathbb{R} \cup \{+\infty\}$, $n \geq 1$, are proper functions, then we say that *the sequence* $\{\varphi_n\}_{n \geq 1}$ *converges to* φ *in the Mosco sense*, denoted by $\varphi_n \xrightarrow{M} \varphi$ if and only if epi $\varphi_n \xrightarrow{M}$ epi φ in $X \times \mathbb{R}$. This is also equivalent to the following two conditions:

(a) for every $x \in X$ and every $x_n \xrightarrow{w} x$ in X, we have
$$\varphi(x) \leq \liminf_{n \to \infty} \varphi_n(x_n);$$

(b) for every $x \in X$ we can find a sequence $x_n \to x$ in X such that
$$\varphi(x) = \lim_{n \to \infty} \varphi_n(x_n);$$

The Mosco convergence of functions is a variational mode of convergence and is related to the pointwise convergence of the corresponding sequence of Moreau-Yosida regularizations. Namely, if X and X^* are locally uniformly convex with Frechet differentiable norms, $\{\varphi_n\}_{n \geq 1} \subseteq \Gamma_0(X)$ and $\varphi \in \Gamma_0(X)$, then:
$$\varphi_n \xrightarrow{M} \varphi$$
if and only if
$$(\varphi_n)_\lambda(x) \to \varphi_\lambda(x) \text{ for all } \lambda > 0 \text{ and all } x \in X.$$

Moreover, if $\{\varphi_n\}_{n \geq 1} \subseteq \Gamma_0(X)$, $\varphi \in \Gamma_0(X)$ and $\varphi_n \xrightarrow{M} \varphi$, then
$$\partial \varphi_n \xrightarrow{K_{ws}} \partial \varphi$$
in the sense that the following two conditions hold:

(α) if $x_{n_k} \xrightarrow{w} x$, in X, $x_{n_k}^* \to x^*$ in X^*, $x_{n_k}^* \in \partial \varphi_{n_k}(x_{n_k})$ with $n_k < n_{k+1}$, $\forall k \geq 1$, then $x^* \in \partial \varphi(x^*)$;

(β) if $(x, x^*) \in Gr\partial\varphi$, then we can find $(x_n, x_n^*) \in Gr\partial\varphi_n$, $n \geq 1$, such that $x_n \xrightarrow{w} x$, in X and $x_n^* \to x^*$, in X^*.

Throughout the remainder of this Chapter, X is assumed to be a reflexive Banach space.

A function $\varphi : X \to \mathbb{R}$ is said to be *locally Lipschitz*, if for every $x \in X$, we can find a neighborhood U of x and a $k_U > 0$ such that
$$|\varphi(y) - \varphi(z)| \leq k_U \|y - z\| \text{ for all } y, z \in U.$$

For such a function φ, we define the *generalized directional derivative*
$$\varphi^0(x; h) = \limsup_{x' \to x, \lambda \to 0^+} \frac{\varphi(x' + \lambda h) - \varphi(x')}{\lambda}, \quad (x, h \in X).$$

It is easy to check that $\varphi^0(x; .)$ is sublinear and continuous, and so it is the support function of a nonempty, w-compact and convex set $\partial \varphi(x) \subseteq X^*$ defined by
$$\partial \varphi(x) = \{x^* \in X^* : \langle x^*, h \rangle \leq \varphi^0(x; h) \text{ for all } h \in X\}.$$

The multifunction $\partial \varphi : X \to 2^{X^*} \setminus \{\emptyset\}$ is called the *generalized subdifferential* of φ. If $\varphi \in C^1(X)$, then φ is locally Lipschitz and
$$\partial \varphi(x) = \{\nabla \varphi(x)\}.$$

If φ is continuous and convex, then from what was said earlier we know that φ is locally Lipschitz and the generalized and convex subdifferentials coincide.

A multifunction $G : X \to 2^{X^*}$ is said to be *upper semicontinuous* (usc for short), if for every closed set $C \subseteq X$,
$$G^-(C) := \{x \in X : G(x) \cap C \neq \varnothing\}$$
is closed in X. We say that the multifunction $G : X \to 2^{X^*}$ *belongs to class* (P), if it is upper semicontinuous with $G(x)$ nonempty, closed and convex for every $x \in X$ and such that
$$G(A) := \bigcup_{x \in A} G(x)$$
is relatively compact in X^* for any bounded subset A of X.

Recall that if $G : D \subseteq X \to 2^{X^*} \setminus \{\varnothing\}$ is an usc multifunction with closed convex values, then by Cellina's approximate selection theorem [9] (see also [25], Theorem 4.41, p.106), for every $\varepsilon > 0$ there exists a continuous function $g_\varepsilon : D \to X^*$ such that
$$g_\varepsilon(x) \in G((x + B_\varepsilon) \cap D) + B_\varepsilon^*, \text{ for all } x \in D$$
and $g_\varepsilon(D) \subseteq \overline{conv}(G(D))$, with
$$B_\varepsilon = \{x \in X : \|x\| < \varepsilon\} \text{ and } B_\varepsilon^* = \{x^* \in X^* : \|x^*\|_* < \varepsilon\}.$$
In particular, if G belongs to class (P), then the approximate selector g_ε is compact.

By the Troyanski renorming theorem (see for example Gasinski-Papageorgiou [18], p.722), we can equivalently renorm X so that both X and X^* are locally uniformly convex with Fréchet differentiable norms.

So, in what follows we assume that X and X^* are locally uniformly convex.

Let U be a bounded open subset of X and let $S : \overline{U} \to X^*$ be a demicontinuous operator of type $(S)_+$. Let $\{X_\alpha\}_{\alpha \in J}$ be the family of all finite dimensional subspaces of X and let S_α be the Galerkin approximation of S with respect to X_α, that is
$$\langle S_\alpha(x), y \rangle_{X_\alpha} = \langle S(x), y \rangle \text{ for all } x \in \overline{U} \cap X_\alpha \text{ and all } y \in X_\alpha.$$
Then, for $x^* \notin S(\partial U)$, $deg_{(S)_+}(S, U, x^*)$ is defined by
$$deg_{(S)_+}(S, U, x^*) = d_B(S_\alpha, U \cap X_\alpha, x^*)$$
for X_α large enough (in the sense of inclusion), where $d_B(.,.,.)$ stands for the classical Brouwer degree.

If X is separable and S is bounded, then we can use only a countable subfamily $\{X_n\}_{n \geq 1}$ of $\{X_\alpha\}_{\alpha \in J}$ such that
$$\overline{\bigcup_{n \geq 1} X_n} = X.$$

Now, let X and X^* be locally uniformly convex reflexive Banach spaces. Let U be a bounded open subset in X, $S : \overline{U} \to X^*$ a bounded, demicontinuous operator of type $(S)_+$ and $A : D(A) \subseteq X \to 2^{X^*} \setminus \{\varnothing\}$ a maximal monotone operator with $0 \in A(0)$.

Then for every $\lambda > 0$, the operator $S + A_\lambda$ is a bounded, demicontinuous operator of type $(S)_+$.

For every $x^* \notin (S + A)(\partial U)$, $deg_0(S + A, U, x^*)$ is defined by
$$deg_0(S + A, U, x^*) = deg_{(S)_+}(S + A_\lambda, U, x^*),$$

for all sufficiently small $\lambda > 0$.

Finally, if everything is as above, and in addition we have a multifunction G in the class (P), then for $x^* \notin (S + A + G)(\partial U)$, $deg(S + A + G, U, x^*)$ is defined by

$$deg(S + A + G, U, x^*) = deg_0(S + A + g_\varepsilon, U, x^*)$$

for $\varepsilon > 0$ small, where g_ε is a continuous ε-approximate selector of G as described earlier.

Recall that in degree theory, the homotopy invariance of the degree map is with respect to a certain class of admissible homotopies. Next we introduce the admissible homotopies for the maps S, A and G.

DEFINITION 1. (a) A one-parameter family $\{S_t\}_{t \in [0,1]}$ of maps from $\overline{U}$ into X^* is said to be a "homotopy of class $(S)_+$", if for any $\{x_n\}_{n \geq 1} \subseteq \overline{U}$ such that $x_n \overset{w}{\to} x$, weakly in X, and for any $\{t_n\}_{n \geq 1} \subseteq [0,1]$ with $t_n \to t$ for which

$$\limsup_{n \to \infty} \langle S_{t_n}(x_n), x_n - x \rangle \leq 0,$$

we have that $x_n \to x$ in X and $S_{t_n}(x_n) \overset{w}{\to} S_t(x)$ in X^*.

(b) A family $\{A^t\}_{t \in [0,1]}$ of maximal monotone maps from X into X^* such that $(0,0) \in GrA^t$ for all $t \in [0,1]$ is said to be a "pseudomonotone homotopy", if it satisfies the following mutually equivalent conditions

(b_1) if $t_n \to t$ in $[0,1]$, $x_n \overset{w}{\to} x$ in X, $x_n^* \overset{w}{\to} x^*$ in X^*, $x_n^* \in A^{t_n}(x_n)$ and

$$\limsup_{n \to \infty} \langle x_n^*, x_n - x \rangle \leq 0,$$

then $(x, x^*) \in GrA^t$ and $\langle x_n^*, x_n \rangle \to \langle x^*, x \rangle$;

(b_2) $(t, x^*) \to \xi(t, x^*) = (A^t + \mathcal{F})^{-1}(x^*)$ is continuous from $[0,1] \times X^*$ into X, where both X and X^* are equipped with their respective norm topologies;

(b_3) for every $x^* \in X^*$, $t \to \xi(t, x^*) = (A^t + \mathcal{F})^{-1}(x^*)$ is continuous from $[0,1]$ into X endowed with the norm topology;

(b_4) if $t_n \to t$ in $[0,1]$, then

$$GrA^t \subseteq s - \liminf_{n \to \infty} GrA^{t_n}.$$

(c) A one-parameter family $\{G_t\}_{t \in [0,1]}$ of multifunctions $G_t : \overline{U} \to 2^{X^*} \setminus \{\varnothing\}$ is said to be a "homotopy of class (P)" if $(t, x) \to G_t(x)$ is usc from $[0,1] \times \overline{U}$ into $2^{X^*} \setminus \{\varnothing\}$, for every $(t, x) \in [0,1] \times \overline{U}$, $G_t(x) \subseteq X^*$ is closed and convex and

$$\overline{\bigcup \{G_t(x) : t \in [0,1], \; x \in \overline{U}\}}$$

is compact in X^*.

REMARK 2. Since for every $\lambda > 0$, $J_\lambda^{A^t}(.)$ and $A_\lambda^t(.)$ are both Lipschitz continuous and for every $(t, x) \in [0,1] \times X$,

$$(\mathcal{F} + A^t)^{-1} \left[\mathcal{F}\left(J_\lambda^{A^t}(x)\right) + A_\lambda^t(x) \right] = J_\lambda^{A^t}(x),$$

we see that equivalently (b_2) and (b_3) can be rewritten as follows:

$(b_2)'$ for every $\lambda > 0$ (equivalently for some $\lambda > 0$), the map $(t, x) \to J_\lambda^{A^t}(x)$ is continuous from $[0,1]$ into X with the strong topology;

$(b_3)'$ for each $\lambda > 0$ (equivalently for some $\lambda > 0$), the map $t \to J_\lambda^{A^t}(x)$ is continuous from $[0,1]$ into X with the strong topology;

Then the homotopy invariance of the degree map "deg", can be formulated as follows:

"If $\{S_t\}_{t \in [0,1]}$ is a homotopy of class $(S)_+$ such that each S_t is bounded, $\{A^t\}_{t \in [0,1]}$ is a pseudomonotone homotopy of maximal monotone operators with $0 \in A^t(0)$ for all $t \in [0,1]$, $\{G_t\}_{t \in [0,1]}$ is a homotopy of class (P) and $x^* : [0,1] \to X^*$ is a continuous map such that

$$x_t^* \notin (S_t + A_t + G_t)(\partial U)$$

for all $t \in [0,1]$, then $\deg(S_t + A_t + G_t, U, x_t^*)$ is independent of $t \in [0,1]$."

Concerning the normalization property of the degree map, we have

$$\deg(\mathcal{F}, U, x^*) = 1 \text{ for all } x^* \in \mathcal{F}(U).$$

The degree map has all the usual properties (such as solution property, additivity of domain property, excision property). Concerning the degree maps $\deg_{(S)_+}$ and $\deg_0$ we refer the reader to Browder [8], while for the degree map deg we refer to Hu-Papageorgiou [24].

Finally let us recall some basic facts about the spectrum of the negative p-Laplacian with Dirichlet boundary conditions. So let $Z \subseteq \mathbb{R}^N$ be a bounded domain with a C^2-boundary ∂Z, let $m \in L^\infty(Z)_+$, $m \neq 0$, and consider the following nonlinear weighted (with weight m) eigenvalue problem

$$\begin{cases} -div\left(\|Dx(z)\|_{\mathbb{R}^N}^{p-2} Dx(z)\right) = \widehat{\lambda} m(z) |x(z)|^{p-2} x(z) \text{ a.e. on } Z, \\ x|_{\partial Z} = 0, \end{cases}$$

where $\widehat{\lambda} \in \mathbb{R}$ and $1 < p < \infty$.

This problem has a smallest (principal) eigenvalue denoted by $\widehat{\lambda}_1(m)$ which is positive, isolated and simple (i.e., the corresponding eigenspace is one-dimensional). Via the Rayleigh quotient, we have a variational characterization of $\widehat{\lambda}_1(m)$, which is

$$(2.1) \qquad \widehat{\lambda}_1(m) = \inf\left\{\frac{\|Dx\|_p^p}{\int_Z m(z)|x(z)|^p dz} : x \in W_0^{1,p}(Z), x \neq 0\right\}$$

where $\|Dx\|_p$ indicates the norm in $L^p(Z, \mathbb{R}^N)$.

In (2.1) the infimum is actually attained at a corresponding eigenfunction u_1. From the nonlinear regularity theory we know that $u_1 \in C_0^1(\overline{Z})$ (see for example Gasinski-Papageorgiou [18], pp.115-116). Moreover, it is clear from (2.1) that u_1 does not change sign. So we may assume that $u_1(z) \geq 0$ for all $z \in Z$.

In fact invoking the nonlinear strict maximum principle of Vazquez [45] (see also Gasinski-Papageorgiou [18], pp.116-117), we have that

$$u_1(z) > 0 \text{ for all } z \in Z \text{ and } \frac{\partial u_1}{\partial n}(z) < 0 \text{ for all } z \in \partial Z,$$

where n is the unit outward normal on ∂Z.

If $u \in W_0^{1,p}(Z)$ is an eigenfunction corresponding to an eigenvalue $\widehat{\lambda} \neq \widehat{\lambda}_1(m)$, then $u \in C_0^1(\overline{Z})$ and u must change sign.

If $m_1, m_2 \in L^\infty(Z)_+$ are two weight functions such that $m_1(z) \leq m_2(z)$ a.e. on Z with strict inequality on a set of positive measure, then we have
$$\widehat{\lambda}_1(m_2) < \widehat{\lambda}_1(m_1),$$
i.e., we have strict monotonicity of the principal eigenvalue $\widehat{\lambda}_1(m)$ with respect to the weight function. If $m \equiv 1$, then we write $\widehat{\lambda}_1(1) =: \lambda_1$. In addition to $\widehat{\lambda}_1 = \widehat{\lambda}_1(m)$, the Liusternik-Schnirelmann theory gives a whole strictly increasing sequence
$$\left\{\widehat{\lambda}_k = \widehat{\lambda}_k(m)\right\}_{k \geq 2} \subseteq (0, +\infty)$$
for which the nonlinear eigenvalue problem has a nontrivial solution.

We have that $\widehat{\lambda}_k \to +\infty$ as $k \to \infty$ and the $\widehat{\lambda}_k$'s are called the "*Liusternik-Schnirelmann eigenvalues*" or "*variational eigenvalues*". If $p = 2$, then these are all the eigenvalues of $(-\triangle, H_0^1(Z))$. If $p \neq 2$, then we do not know if this is the case. However, if we set
$$\widehat{\lambda}_2^* = \inf\left\{\widehat{\lambda} : \widehat{\lambda} > \widehat{\lambda}_1(m), \widehat{\lambda} \text{ eigenvalue of } \left(-\triangle_p, W_0^{1,p}(Z)\right) \text{ with weight } m\right\}$$
then $\widehat{\lambda}_2 = \widehat{\lambda}_2^*$. In other words, the second eigenvalue and the second variational eigenvalue coincide.

Finally, if $m_1, m_2 \in L^\infty(Z)_+$ and $m_1(z) < m_2(z)$ a.e. on Z, then we have
$$\widehat{\lambda}_2(m_2) < \widehat{\lambda}_2(m_1).$$

For details we refer to Anane [**3**], Anane-Tsouli [**4**] and Denkowski-Migorski-Papageorgiou [**12**].

CHAPTER 3

Degree Theoretic Results

In this Chapter we prove an extension of Amann's theorem to operators which are the sum of a generalized subdifferential and a convex subdifferential. To do this, we will need a general result about the degree map for such operators. For this purpose, let $Z \subseteq \mathbb{R}^N$ be a bounded open set with a C^2−boundary ∂Z and introduce an integrand $j_0 : Z \times \mathbb{R} \to \mathbb{R}$ which satisfies the following hypotheses:

(H_{j_0}) : The function $j_0 : Z \times \mathbb{R} \to \mathbb{R}$ is such that $j_0(.,0) \in L^\infty(Z)$ and
   (i) for all $x \in \mathbb{R}$, $z \to j_0(z,x)$ is measurable;
   (ii) for almost all $z \in Z$, $x \to j_0(z,x)$ is locally Lipschitz;
   (iii) there exist $a \in L^\infty(Z)_+$ and $c > 0$ such that for almost all $z \in Z$, all $x \in \mathbb{R}$ and all $u \in \partial j_0(z,x)$, we have
$$|u| \leq a(z) + c|x|^{p-1} \quad (1 < p < \infty).$$

We now define the integral functional $\widehat{J}_0 : L^p(Z) \to \mathbb{R}$ by

(3.1) $$\widehat{J}_0(x) = \int_Z j_0(z, x(z))\, dz \text{ for all } x \in L^p(Z).$$

We know that under hypotheses (H_{j_0}), the integral functional $\widehat{J}_0$ is Lipschitz continuous on bounded sets, hence it is locally Lipschitz (see Gasinski-Papageorgiou [18], p.59).

Let $N_0 : L^p(Z) \to 2^{L^q(Z)}$ $\left(\frac{1}{p} + \frac{1}{q} = 1\right)$ be defined by

$$N_0(x) = \{u^* \in L^q(Z) : u^*(z) \in \partial j_0(z, x(z)) \text{ a.e. on } Z\},\ x \in L^p(Z).$$

PROPOSITION 3. *If hypotheses (H_{j_0}) hold, then N_0 has nonempty, weakly compact and convex values in $L^q(Z)$ and it is usc from $L^p(Z)$ with the norm topology into $L^q(Z)$ with the weak topology.*

PROOF. Clearly N_0 has weakly compact and convex values. What is not immediately clear is that they are nonempty. To show this, let $x \in L^p(Z)$ and let $\{s_n\}_{n \geq 1}$ be simple functions such that $s_n(z) \to x(z)$ a.e. on Z and

$$|s_n(z)| \leq |x(z)| \text{ a.e. on } Z.$$

Since for every $x \in \mathbb{R}$, $z \to \partial j_0(z,x)$ is graph measurable, then for every $n \geq 1$, $z \to \partial j_0(z, s_n(z))$ is graph measurable, too. By applying the Yankov-von Neumann-Aumann selection theorem (see Hu-Papageorgiou [25], p.158), we obtain $f_n : Z \to \mathbb{R}$, $n \geq 1$, a Lebesgue measurable function, such that

$$f_n(z) \in \partial j_0(z, s_n(z)) \text{ a.e. on } Z.$$

By virtue of hypothesis (H_{j_0}) (iii), $\{f_n\}_{n \geq 1} \subseteq L^q(Z)$ is bounded so we may assume that $f_n \overset{w}{\to} f$ in $L^q(Z)$. Since $\partial j_0(z,.)$ is usc with convex values, by Mazur's lemma,

we have
$$f(z) \in \partial j_0(z, x(z)) \text{ a.e. on } Z,$$
i.e., $f \in N_0(x)$. Since the weak topology on bounded subsets of $L^q(Z)$ is metrizable, to show the desired upper semicontinuity of N_0, it suffices to show that GrN_0 is sequentially closed in $L^p(Z) \times L^q(Z)_w$ (where $L^q(Z)_w$ denotes the Lebesgue space $L^q(Z)$ equipped with the weak topology). So suppose $x_n \to x$ in $L^p(Z)$, $f_n \overset{w}{\to} f$ in $L^q(Z)$ and $f_n(z) \in \partial j_0(z, x_n(z))$ a.e. on Z, for all $n \geq 1$. We may assume that $x_n(z) \to x(z)$ a.e. on Z. So, as above, by Mazur's lemma and since $\partial j_0(z, .)$ is usc with convex values, we obtain $f(z) \in \partial j_0(z, x(z))$ a.e. on Z, which proves the desired closedness of the graph of N_0. Therefore N_0 is usc from $L^p(Z)$ into $L^q(Z)_w$. □

Suppose X is a reflexive Banach space which is embedded compactly and densely in $L^p(Z)$. Then $L^q(Z)$ is embedded compactly and densely in X^*, and so we have:

COROLLARY 4. *If hypotheses* (H_{j_0}) *hold, then* $N_0 : X \to 2^{X^*} \setminus \{\varnothing\}$ *is a multifunction of class* (P).

So if $\theta \in C^1(X)$ and $\nabla \theta = \widehat{A}$ is a bounded, $(S)_+$ operator, $J_0 = \widehat{J_0}|_X$ and $\widehat{\varphi} = \theta + J_0$, then we can speak about the degree of $\partial \widehat{\varphi} = \widehat{A} + N_0$ with
$$N_0 = \partial J_0 = \partial \widehat{J_0}$$
(see Gasinski-Papageorgiou [**18**], p. 56 for the last equality).

If $U \subseteq X$ is a nonempty open set, let $\{X_\beta\}_{\beta \in S}$ be the collection of all finite dimensional subspaces of X such that $U \cap X_\beta \neq \varnothing$ for all $\beta \in S$, ordered by inclusion, namely
$$\beta_1 \leq \beta_2 \text{ if and only if } X_{\beta_1} \subseteq X_{\beta_2}.$$
Let $i_\beta : X_\beta \to X$ be the embedding operator of X_β into X. We set
$$\widehat{\varphi}_\beta := \widehat{\varphi} \circ i_\beta,$$
$$\theta_\beta := \theta \circ i_\beta,$$
and
(3.2) $$J_\beta^0 := J_0 \circ i_\beta.$$
where, as before, $\theta \in C^1(X)$ and $J_0 = \widehat{J_0}|_X$, with $\widehat{J_0}$ given by (3.1).

PROPOSITION 5. *If X is a reflexive Banach space which is embedded compactly and densely in $L^p(Z)$ $(1 < p < \infty)$, $U \subseteq X$ is a nonempty open set, $\widehat{\varphi} = \theta + J_0 : X \to \mathbb{R}$ is locally Lipschitz with $\theta \in C^1(X)$, $\nabla \theta = \widehat{A}$ is a bounded, $(S)_+$ operator, $J_0 = \widehat{J_0}|_X$ with $\widehat{J_0}$ given by (3.1) and $x_0 \in U$, $\xi, \mu, r \in \mathbb{R}$, $\xi < \mu$ satisfy*
  (i) $x_0 \in V := \{\widehat{\varphi} < \mu\} \cap U$ *and* $\{\widehat{\varphi} \leq \mu\} \cap \overline{U}$ *is a bounded subset of* U,
  (ii) *if* $x \in \{\widehat{\varphi} \leq \xi\} \cap U$, *then* $tx_0 + (1-t)x \in V$ *for all* $t \in [0, 1]$,
  (iii) $0 \notin \partial \widehat{\varphi}(x)$ *for all* $x \in \{\xi \leq \widehat{\varphi} \leq \mu\} \cap \overline{U}$,
then
$$\deg(\partial \widehat{\varphi}, V, 0) = 1.$$

PROOF. Let $\{X_\beta\}_{\beta \in S}$ be the collection of all finite dimensional subspaces of X such that $U \cap X_\beta \neq \varnothing$ for all $\beta \in S$, ordered by inclusion, namely

$$\beta_1 \leq \beta_2 \text{ if and only if } X_{\beta_1} \subseteq X_{\beta_2}.$$

Let $i_\beta : X_\beta \to X$ be the embedding operator of X_β into X. We set

$$\widehat{\varphi}_\beta := \widehat{\varphi} \circ i_\beta.$$

Then $\widehat{\varphi}_\beta : X_\beta \to \mathbb{R}$ and from the nonsmooth chain rule (see Gasinski-Papageorgiou [**18**], pp.54-55), we have

(3.3) $\qquad \partial \widehat{\varphi}_\beta(x) \subseteq i_\beta^* \partial \widehat{\varphi}_\beta(i_\beta(x)) = p_\beta \partial \widehat{\varphi}(x)$ for all $x \in X_\beta$,

where $p_\beta = i_\beta^* : X^* \to X_\beta^*$ is the projection of X^* onto X_β^* (which exists since X_β^* is finite dimensional). In (3.3) above we consider the generalized subdifferentials of the locally Lipschitz functions $\widehat{\varphi}$ and $\widehat{\varphi}_\beta$. *Claim:* There exists $\beta_0 \in S$ such that for all $\beta \geq \beta_0$ in S, we have $0 \notin \partial \widehat{\varphi}_\beta(x)$ for all $x \in \{\xi \leq \widehat{\varphi} \leq \mu\} \cap \overline{U}$. Let

$$W_{\xi,\mu} := \{\xi \leq \widehat{\varphi} \leq \mu\} \cap \overline{U} = \{x \in \overline{U} : \xi \leq \widehat{\varphi}(x) \leq \mu\}$$

We proceed by contradiction. So suppose that the *Claim* is not true. Then for every $\beta_0 \in S$ we can find $\beta \geq \beta_0$ and $x_\beta \in W_{\xi,\mu}$ such that $0 \in \partial \widehat{\varphi}_\beta(x_\beta)$. Because of (3.3), we have $p_\beta x_\beta^* = 0$ for some $x_\beta^* \in \partial \widehat{\varphi}(x_\beta)$, hence

$$\langle p_\beta x_\beta^*, v \rangle_{X_\beta} = \langle x_\beta^*, i_\beta(v) \rangle \text{ for all } v \in X_\beta \text{ (recall that } p_\beta^* = i_\beta \text{)}$$

For every $\beta \in S$, we introduce the set

$$E_\beta = \{x \in W_{\xi,\mu} : \exists x^* \in \partial \widehat{\varphi}(x) \text{ with } \langle x^*, x \rangle = 0 \text{ and } \langle x^*, v \rangle = 0, \forall v \in X_\beta\}.$$

Evidently if $\beta_1 \leq \beta_2$ in S, then $E_{\beta_1} \subseteq E_{\beta_2}$. Hence, by virtue of the previous argument, we have that $E_\beta \neq \varnothing$ for all $\beta \in S$. So if S_f is a finite subset of S, then

$$\bigcap_{\beta \in S_f} E_\beta \neq \varnothing.$$

By hypothesis (i) the set $\{\widehat{\varphi} \leq \mu\} \cap \overline{U} = \{x \in \overline{U} : \widehat{\varphi}(x) \leq \mu\}$ is bounded in X. Because X is a reflexive Banach space, it follows that $E_\beta \subseteq X$ is relatively weakly compact. So from the finite intersection property, we infer that

$$\bigcap_{\beta \in S} \overline{E_\beta}^w \neq \varnothing.$$

Let $x \in \bigcap_{\beta \in S} \overline{E_\beta}^w$ and consider any $u \in X$. We chose $\beta \in S$ such that $x, u \in X_\beta$. We can find $\{x_n\}_{n \geq 1} \subseteq E_\beta$ $(\beta \in S)$ such that $x_n \xrightarrow{w} x$ in X (see Denkowski-Migorski-Papageorgiou [**13**], p.306). From the definition of the set E_β, we know that there exists $x_n^* \in \partial \widehat{\varphi}(x_n)$ such that

$$\langle x_n^*, x_n \rangle = 0 \text{ and } \langle x_n^*, v \rangle = 0, \forall v \in X_\beta.$$

Since $x \in X_\beta$, we have $\langle x_n^*, x_n - x \rangle = 0$, hence

(3.4) $\qquad \left\langle \widehat{A}(x_n) + u_n^*, x_n - x \right\rangle = 0$

with

$$u_n^* \in \partial J_0(x_n) = \partial \widehat{J_0}(x_n) \subseteq L^q(Z) \, (\frac{1}{p} + \frac{1}{q} = 1)$$

and

$$u_n^*(z) \in \partial j_0(z, x_n(z)) \text{ a.e. on } Z$$

(see Gasinski-Papageorgiou [**18**], pp.56 and 59). Because X is embedded compactly in $L^p(Z)$, we may assume that $x_n \to x$ in $L^p(Z)$, while from hypothesis (H_{j_0}) (iii) we see that $\{u_n^*\}_{n \geq 1} \subseteq L^q(Z)$ is bounded. So,
$$\langle u_n^*, x_n - x \rangle = \int_Z u_n^*(z)(x_n - x)(z)\,dz \to 0 \text{ as } n \to \infty.$$
Then from (3.4) it follows that
$$(3.5) \qquad \lim_{n \to \infty} \left\langle \widehat{A}(x_n), x_n - x \right\rangle = 0.$$
Since by hypothesis $\widehat{A}$ is an operator of type $(S)_+$, from (3.5) we infer that
$$x_n \to x \text{ in } X.$$
Note that the set
$$\{\xi \leq \widehat{\varphi} \leq \mu\} \cap U = \{x \in U : \xi \leq \widehat{\varphi}(x) \leq \mu\}$$
is closed in U with the relative strong topology. Since
$$\{x_n\}_{n \geq 1} \subseteq \{\xi \leq \widehat{\varphi} \leq \mu\} \cap U,$$
we infer that
$$x \in \{\xi \leq \widehat{\varphi} \leq \mu\} \cap U.$$
Moreover, since $x_n \to x$ in X and $x_n^* \in \partial\widehat{\varphi}(x_n)$ for all $n \geq 1$, it follows that $\{x_n^*\}_{n \geq 1} \subseteq X^*$ is bounded. So, by the Eberlein-Smulian theorem and passing to a suitable subsequence if necessary, we may assume that
$$x_n^* \xrightarrow{w} x^* \text{ in } X^* \text{ as } n \to \infty.$$
From the definition of the generalized subdifferential, we have
$$\langle x_n^*, h \rangle \leq \varphi^0(x_n; h) \text{ for all } h \in X \text{ and all } n \geq 1.$$
Passing to the limit as $n \to \infty$ and using the fact that $(x, h) \to \varphi^0(x; h)$ is upper semicontinuous on $X \times X$ (see Gasinski-Papageorgiou [**18**], p.49), we obtain $\langle x^*, h \rangle \leq \varphi^0(x; h)$ for all $h \in X$, hence
$$x^* \in \partial\widehat{\varphi}(x).$$
Also note that
$$(3.6) \qquad\qquad\qquad \langle x^*, u \rangle = 0.$$
Recalling that $u \in X$ was arbitrary, we infer that
$$x^* = 0.$$
On the other hand, we know that $x \in \{\xi \leq \widehat{\varphi} \leq \mu\} \cap U$. Since $x^* \in \partial\widehat{\varphi}(x)$, from hypothesis (iii) it follows that
$$x^* \neq 0,$$
a contradiction. This proves the *Claim*. Next let $\beta \in S$ be such that $\beta \geq \beta_0$ (with $\beta_0 \in S$ as in the Claim) and let $x_0 \in X_\beta$. Then $V_\beta := \{\widehat{\varphi}_\beta < \mu\} \cap U$ is nonempty and
$$\{\widehat{\varphi}_\beta \leq \mu\} \cap \overline{U} \subseteq U \cap X_\beta = U_\beta$$
is bounded (see hypothesis (i)). If $x \in \{\widehat{\varphi}_\beta \leq \xi\}$, then $tx_0 + (1-t)x \in V_\beta$ (see hypothesis (ii)) and $0 \notin \partial\widehat{\varphi}(x)$ for all $x \in \{\xi \leq \widehat{\varphi}_\beta \leq \mu\} \cap U$ (see the Claim). Given $\delta > 0$, we consider the the multifunction $\Gamma_\delta : Z \to 2^{C(\mathbb{R},\mathbb{R})}$ defined by
$$\Gamma_\delta(z) = \{g \in C(\mathbb{R}, \mathbb{R}) : g(r) \in \partial j_0(z, r + B_\delta) + B_\delta \text{ for all } r \in \mathbb{R}\}$$

where $B_\delta = (-\delta, \delta)$. Recall that the generalized subdifferential $x \to \partial j_0(z, x)$ is upper semicontinuous with nonempty, compact and convex values (see for example Gasinski-Papageorgiou [**18**], p.50). So, by Cellina's approximate selection theorem for upper semicontinuous multifunctions (see Chapter 2), we know that $\Gamma_\delta(z) \neq \varnothing$ for almost all $z \in Z$. By redefining the multifunction Γ_δ on the exceptional Lebesgue null set in Z, we may assume without loss of generality that $\Gamma_\delta(z) \neq \varnothing$ for all $z \in Z$. We set
$$\widehat{F}_0(z, x) := \partial j_0(z, r + \overline{B}_\delta) + \overline{B}_\delta,$$
where $\overline{B}_\delta = [-\delta, \delta]$. Since the multifunction $x \to \partial j_0(z, x)$ is usc, so is the multifunction $x \to \widehat{F}_0(z, x)$ and, for every $(z, x) \in Z \times \mathbb{R}$, $\widehat{F}_0(z, x)$ is a bounded, closed interval in $\mathbb{R}$ (see Hu-Papageorgiou [**25**], pp.42-43). If by $\sigma\left(.; \widehat{F}_0(z, x)\right)$ we denote the support function of $\widehat{F}_0(z, x)$, then for any $h \in \mathbb{R}$ we have

(3.7) $\quad \sigma\left(h; \widehat{F}_0(z, x)\right) = \max\left\{\sigma(h; \partial j_0(z, r + v)) : v \in \overline{B}_\delta\right\} + \delta.$

For fixed $r \in \mathbb{R}$ and every $h \in \mathbb{R}$, we have

$$j_0^0(z, r + v; h) = \limsup_{w \to v, \lambda \downarrow 0} \frac{j_0(z, r + w + \lambda h) - j_0(z, r + w)}{\lambda}$$

(3.8) $\qquad = \inf_{m \geq 1} \sup_{t,s \in Q \cap \left(-\frac{1}{m}, \frac{1}{m}\right)} \frac{j_0(z, r + t + sh) - j_0(z, r + t)}{s}.$

Recall that $(z, x) \to j_0(z, x)$ is a Carathéodory function (i.e., measurable in $z \in Z$ and continuous in $x \in \mathbb{R}$), hence it is jointly measurable. So, from (3.8) it follows that
$$(z, v) \to j_0^0(z, r + v; h) = \sigma(h; \partial j_0(z, r + v))$$
is jointly measurable on $Z \times \mathbb{R}$. Then from (3.7) we infer that for fixed $r \in \mathbb{R}$, $z \to \sigma\left(h; \widehat{F}_0(z, r)\right)$ is Lebesgue measurable (see Hu-Papageorgiou [**25**], p.161) and so $z \to \widehat{F}_0(z, r)$ is a Lebesgue measurable multifunction (see Hu-Papageorgiou [**25**], p.166). We have
$$Gr\Gamma_\delta = \left\{(z, g) \in Z \times C(\mathbb{R}, \mathbb{R}) : d\left(g(r), \widehat{F}_0(z, r)\right) = 0 \text{ for all } r \in \mathbb{R}\right\}.$$
Let $\{r_n\}_{n \geq 1}$ be an enumeration of the rationals. Note that the upper semicontinuity of $\widehat{F}_0(z, .)$ implies that $r \to d\left(g(r), \widehat{F}_0(z, r)\right)$ is a lower semicontinuous, $\mathbb{R}_+$ - valued function (see Hu-Papageorgiou [**25**], p.61). Hence

(3.9) $\qquad Gr\Gamma_\delta = \bigcap_{n \geq 1} \left\{(z, g) \in Z \times C(\mathbb{R}, \mathbb{R}) : d\left(g(r_n), \widehat{F}_0(z, r_n)\right) = 0\right\}.$

For every $n \geq 1$, the function $(z, g) \to d\left(g(r_n), \widehat{F}_0(z, r_n)\right)$ is a Carathéo-dory function on $Z \times C(\mathbb{R}, \mathbb{R})$. Since $C(\mathbb{R}, \mathbb{R})$ is a separable Fréchet space, we infer that $(z, g) \to d\left(g(r_n), \widehat{F}_0(z, r_n)\right)$ is jointly measurable. Hence from (3.9) it follows that $Gr\Gamma_\delta \in \mathcal{L}(Z) \times \mathcal{B}(C(\mathbb{R}, \mathbb{R}))$, where $\mathcal{L}(Z)$ is the Lebesgue σ-field of Z and $\mathcal{B}(C(\mathbb{R}, \mathbb{R}))$ is the Borel σ-field of $C(\mathbb{R}, \mathbb{R})$. We apply the Yankov-von Neumann-Aumann selection theorem (see Hu-Papageorgiou [**25**], p.158) and obtain a Lebesgue measurable map $\gamma_\delta : Z \to C(\mathbb{R}, \mathbb{R})$ such that
$$\gamma_\delta(z) \in \Gamma_\delta(z) \text{ for all } z \in Z.$$

We set
$$g_\delta(z,x) := \gamma_\delta(z)(x) \text{ for all } (z,x) \in Z \times \mathbb{R}.$$
Evidently $g_\delta(.,.)$ is a Carathéodory function and we have
$$g_\delta(z,x) \in \partial j_0(z, x + \overline{B}_\delta) + \overline{B}_\delta \text{ for a.a. } z \in Z, \text{ all } x \in \mathbb{R},$$
where $\overline{B}_\delta = [-\delta, \delta]$. By $\widehat{g}_\delta : L^p(Z) \to L^q(Z)$ we denote the Nemitsky operator corresponding to the Carathéodory function $g_\delta(.,.)$, i.e.,
$$\widehat{g}_\delta(x)(.) := g(., x(.)) \text{ for all } x \in L^p(Z).$$
We set
$$G_\delta(z,x) := \int_0^x g_\delta(z,r)\, dr \text{ for all } (z,x) \in Z \times \mathbb{R},$$
i.e., $G_\delta(z,x)$ is the potential function corresponding to g_δ. We introduce the functional $T_\delta : X_\beta \to \mathbb{R}$ defined by
$$T_\delta(x) = \int_Z G_\delta(z, x(z))\, dz + J_\beta^0(x_0) - \int_Z G_\delta(z, x_0(z))\, dz \; \forall x \in X_\beta,$$
where J_β^0 is defined by (3.2). Evidently $T_\delta(x_0) = J_\beta^0(x_0)$ and moreover, $T_\delta \in C^1(X_\beta)$ and $\nabla T_\delta(x) = \widehat{g}_\delta(x)$ for all $x \in X_\beta$ (see Denkowski-Migorski-Papageorgiou [12], p.89). Recall that $\beta \geq \beta_0$ is such that $x_0 \in X_\beta$. The space $X_\beta \subseteq X \subseteq L^p(Z)$ is finite dimensional and so all the norms are equivalent. In particular, if we use the $L^1(Z)$−norm on X_β, from the properties of the function $g_\delta(z,x)$, we see that

(3.10) $\quad \widehat{g}_\delta(x) \in \partial J_\beta^0(x + B_{\delta_1}) + B_{\delta_1}$ for all $x \in \overline{U}_\beta$ with $U_\beta = U \cap X_\beta$,

where
$$B_{\delta_1} = \{x \in X_\beta : \|x\| < \delta_1\}, \delta_1 > \frac{\delta}{c} |Z|_N,$$
with $c > 0$ such that $c\|.\| \leq \|.\|_1$ and $|.|_N$ is the Lebesgue measure on $\mathbb{R}^N$. Since by hypothesis (i), $\{\widehat{\varphi} \leq \mu\} \cap U$ is bounded, without any loss of generality we may assume that U is bounded in X, hence $\overline{U}_\beta = \overline{U \cap X_\beta}$ is bounded and so compact in X_β. The generalized subdifferential $x \to \partial J_\beta^0(x)$ is usc from X_β into $2^{X_\beta} \setminus \{\varnothing\}$ and has compact and convex values. So it is h-usc (see Hu-Papageorgiou [25], pp.59-60) and because $\overline{U}_\beta$ is compact in X_β, the h-upper semicontinuity of $x \to \partial J_\beta^0(x)$ is uniform on $\overline{U}_\beta$. Therefore given $\varepsilon > 0$, we can find $\delta_1 = \delta_1(\varepsilon) \in (0, \varepsilon)$ such that
$$\partial J_\beta^0(x + B_{\delta_1}) \subseteq \partial J_\beta^0(x) + B_\varepsilon \text{ for all } x \in \overline{U}_\beta,$$
hence

(3.11) $\quad \widehat{g}_\delta(x) \in \partial J_\beta^0(x) + B_{2\varepsilon}$ for all $x \in \overline{U}_\beta$

(see (3.10) and recall that $0 < \delta_1 < \varepsilon$). From the nonsmooth mean value theorem (see Gasinski-Papageorgiou [18], p.53), we have, for each $x \in \overline{U}_\beta$

(3.12) $\quad J_\beta^0(x) - T_\delta(x) \in \langle \partial J_\beta^0(\lambda x + (1-\lambda)x_0) - \widehat{g}_\delta(\lambda x + (1-\lambda)x_0), x - x_0 \rangle_{X_\beta}$

for some $\lambda \in (0,1)$. Because of (3.11), for every $u_\lambda \in \partial J_\beta^0(\lambda x + (1-\lambda)x_0)$, we have

(3.13) $\quad \left| \langle u_\lambda - \widehat{g}_\delta(\lambda x + (1-\lambda)x_0), x - x_0 \rangle_{X_\beta} \right| \leq 2\varepsilon \|x - x_0\|.$

We set
$$k_\beta = \max\{\|x - x_0\| : x \in \overline{U}_\beta\}$$

and
(3.14) $$\widehat{\mu} = \mu + 2\varepsilon k_\beta.$$
If $x \in V_\beta := \{\widehat{\varphi}_\beta < \mu\} \cap U_\beta$, then from (3.12), (3.13) and (3.14) we have
$$\theta_\beta(x) + T_\delta(x) \le \theta_\beta(x) + J_\beta^0(x) + 2\varepsilon \|x - x_0\| < \mu + 2\varepsilon \|x - x_0\| < \widehat{\mu},$$
hence
(3.15) $$x \in \widehat{V}_\beta := \{\widehat{\varphi}_\beta < \widehat{\mu}\} \cap U_\beta, \text{ i.e., } V_\beta \subseteq \widehat{V}_\beta.$$
Moreover, since by hypothesis (i) $x_0 \in V \cap X_\beta = V_\beta$, we have
(3.16) $$x_0 \in \widehat{V}_\beta.$$
Next, let $x \in \widehat{V}_\beta$. Using (3.12), (3.13) and (3.14), we have
$$\mu + 2\varepsilon k_\beta = \widehat{\mu} > \theta_\beta(x) + T_\delta(x) \ge \widehat{\varphi}_\beta(x) - 2\varepsilon \|x - x_0\|$$
$$\ge \widehat{\varphi}_\beta(x) - 2\varepsilon k_\beta,$$
hence
$$\widehat{\varphi}_\beta(x) < \mu + 4\varepsilon k_\beta,$$
therefore
(3.17) $$\widehat{V}_\beta \subseteq \{\widehat{\varphi}_\beta < \mu + 4\varepsilon k_\beta\} \cap U_\beta.$$
Since the family $\{\{\widehat{\varphi}_\beta < \mu + 4\varepsilon k_\beta\} \cap U_\beta\}_{\varepsilon > 0}$ of subsets of U_β is decreasing in $\varepsilon > 0$ we conclude that
(3.18) $$\{\widehat{\varphi}_\beta < \mu + 4\varepsilon k_\beta\} \cap U_\beta \overset{K}{\to} \{\widehat{\varphi}_\beta \le \mu\} \cap U_\beta \text{ as } \varepsilon \downarrow 0.$$
(see Hu-Papageorgiou [**25**], p.665). Moreover, we have
$$\overline{V}_\beta = \overline{\{\widehat{\varphi}_\beta < \mu\} \cap U_\beta} \subseteq \{\widehat{\varphi}_\beta \le \mu\} \cap \overline{U}_\beta \subseteq U_\beta.$$
The finite dimensional Banach space X_β is normal. So, we can find an open subset $W \subseteq X_\beta$ such that
$$\overline{V}_\beta \subseteq W \subseteq \overline{W} \subseteq U_\beta.$$
Because of (3.18), for $\varepsilon > 0$ small we have
$$\{\widehat{\varphi}_\beta < \mu + 4\varepsilon k_\beta\} \cap U_\beta \subseteq W \subseteq \overline{W} \subseteq U_\beta,$$
hence
$$\widehat{V}_\beta \subseteq W \subseteq \overline{W} \subseteq U_\beta,$$
(see (3.17)), therefore
(3.19) $$\overline{\widehat{V}_\beta} \subseteq U_\beta.$$
With k_β as above, let
(3.20) $$\widehat{\xi} = \xi + 2\varepsilon k_\beta.$$
Ler $x \in \left\{\theta_\beta + T_\delta \le \widehat{\xi}\right\} \cap U_\beta$. From (3.12), (3.13), (3.14) we have
$$\widehat{\varphi}_\beta(x) - 2\varepsilon \|x - x_0\| \le \theta_\beta(x) + T_\delta(x)$$
$$\le \widehat{\xi} = \xi + 2\varepsilon k_\beta,$$
hence
$$\widehat{\varphi}_\beta(x) \le \xi + 4\varepsilon k_\beta,$$

therefore

(3.21) $$\left\{\theta_\beta + T_\delta \leq \widehat{\xi}\right\} \cap U_\beta \subseteq \{\widehat{\varphi}_\beta \leq \xi + 4\varepsilon k_\beta\} \cap U_\beta.$$

Since $\varepsilon > 0$ was arbitrary, we infer that

(3.22) $$\left\{\theta_\beta + T_\delta \leq \widehat{\xi}\right\} \cap U_\beta \subseteq \{\widehat{\varphi}_\beta \leq \xi\} \cap U_\beta.$$

So if $x \in \left\{\theta_\beta + T_\delta \leq \widehat{\xi}\right\} \cap U_\beta$, then from (3.22) and hypothesis (ii), we have

(3.23) $$tx_0 + (1-t)x \in V_\beta \subseteq \widehat{V}_\beta$$

(see (3.15)). Finally recall that

$$\widehat{g}_\delta(x) \in \partial J_\beta^0(x) + B_{2\varepsilon} \text{ for all } x \in \overline{U}_\beta,$$

hence

$$\widehat{A}_\beta(x) + \widehat{g}_\delta(x) \in \widehat{A}_\beta(x) + \partial J_\beta^0(x) + B_{2\varepsilon} \text{ for all } x \in \overline{U}_\beta.$$

Since $\widehat{A}_\beta = \nabla \theta_\beta$, we get

(3.24) $$\nabla(\theta_\beta + T_\delta)(x) \in \partial \widehat{\varphi}_\beta(x) + B_{2\varepsilon} \text{ for all } x \in \overline{U}_\beta.$$

Set $C_\beta = \{\xi \leq \widehat{\varphi}_\beta \leq \mu\} \cap \overline{U}_\beta$. This set is compact in X_β. Recall that the generalized subdifferential

$$\partial \widehat{\varphi}_\beta : X_\beta \to 2^{X_\beta^*} \setminus \{\varnothing\}$$

is upper semicontinuous with compact, convex values. So, it follows that $\partial \widehat{\varphi}_\beta(C_\beta)$ is compact in X_β^* (see Hu-Papageorgiou [25], p.42) and because of hypothesis (iii)

$$0 \notin \partial \widehat{\varphi}_\beta(C_\beta).$$

Set $\eta := d(0, \partial \widehat{\varphi}_\beta(C_\beta))$. Then $\eta > 0$. If for $\varepsilon' > 0$ we define

$$\overline{(C_\beta)}_{\varepsilon'} = \{x \in \overline{U}_\beta : d(x, C_\beta) \leq \varepsilon'\},$$

then we have

$$\partial \widehat{\varphi}_\beta \left(\overline{(C_\beta)}_{\varepsilon'}\right) \xrightarrow{K} \partial \widehat{\varphi}_\beta(C_\beta) \text{ as } \varepsilon' \to 0.$$

Therefore for $\varepsilon' > 0$ small, we have

$$\partial \widehat{\varphi}_\beta \left(\overline{(C_\beta)}_{\varepsilon'}\right) \subseteq [\partial \widehat{\varphi}_\beta(C_\beta)]_{\frac{\eta}{2}}$$

where

$$[\partial \widehat{\varphi}_\beta(C_\beta)]_{\frac{\eta}{2}} = \left\{x^* \in X_\beta^* : d(x^*, \partial \widehat{\varphi}_\beta(C_\beta)) < \frac{\eta}{2}\right\}.$$

It follows that

(3.25) $$0 \notin \partial \widehat{\varphi}_\beta \left(\overline{(C_\beta)}_{\varepsilon'}\right).$$

From the choices of $\widehat{\xi}$ and $\widehat{\mu}$ (see (3.14) and (3.20), respectively), we see that if $x \in \left\{\widehat{\xi} \leq \theta_\beta + T_\delta \leq \widehat{\mu}\right\}$, then

$$x \in \{\xi \leq \widehat{\varphi}_\beta \leq \mu + 4\varepsilon k_\beta\}.$$

If we choose $0 < \varepsilon < \varepsilon'$ small, then

(3.26) $$\{\xi \leq \widehat{\varphi}_\beta \leq \mu + 4\varepsilon k_\beta\} \subseteq \overline{(C_\beta)}_{\varepsilon'}.$$

So, from (3.24), (3.25) and (3.26), we conclude that

(3.27) $$\nabla(\theta_\beta + T_\delta)(x) \neq 0 \text{ for all } x \in \left\{\widehat{\xi} \leq \theta_\beta + T_\delta \leq \widehat{\mu}\right\} \cap \overline{U}_\beta.$$

Because of (3.16), (3.19), (3.23) and (3.27) we can apply the Theorem of Amann [1]. Note that in the result of Amann, instead of (ii) employed here, it is assumed that
$$\{\theta \leq \mu\} \subseteq \overline{B_r(x_0)} \subseteq U \text{ for some } r > 0,$$
where $\theta \in C^1(U)$ is a compact vector field. This condition is used in Step (iv) in the proof provided by Amann, in order to show that the homotopy $k(\sigma, x)$, $(\sigma, x) \in [0, 1] \times \overline{V}$ (we use the notation of Amann [1], p.595) is admissible on V, namely $k(\sigma, x) \neq 0$ for all $(\sigma, x) \in [0, 1] \times \partial V$. However, it can be easily checked that the admissibility of the homotopy K is still valid if we adopt the weaker hypothesis (ii). This was first observed by Kobayashi and Otani [27]. Applying the Theorem of Amann [1], we obtain
$$d_B\left(\nabla \theta_\beta + \widehat{g}_\delta, \widehat{V}_\beta, 0\right) = 1 \text{ for all } \beta \geq \beta_0.$$

But then from the definition of the generalized Browder degree (see Chapter 2) we conclude that
$$\deg(\partial \widehat{\varphi}, V, 0) = 1.$$
This completes the proof of the proposition. $\square$

Next we will derive some useful properties of the Moreau-Yosida regularization of a proper, convex and lower semicontinuous function. These properties will be used in the proof of the theorem of this Chapter.

So, let X be a reflexive Banach space such that X and X^* are both locally uniformly convex with Fréchet differentiable norms.

LEMMA 6. *If $\psi \in \Gamma_0(X)$ and $\lambda, \mu > 0$, then*
$$(\psi_\lambda)_\mu = \psi_{\lambda + \mu}.$$

PROOF. From the definition of the Moreau-Yosida regularization, for every $x \in X$, we have
$$(\psi_\lambda)_\mu (x) = \inf_{y \in X}\left[\psi_\lambda(y) + \frac{1}{2\mu}\|x - y\|^2\right]$$
$$= \inf_{y \in X} \inf_{v \in X}\left[\psi(v) + \frac{1}{2\lambda}\|y - v\|^2 + \frac{1}{2\mu}\|x - y\|^2\right]$$
$$= \inf_{v \in X}\left[\psi(v) + \inf_{y \in X}\left[\frac{1}{2\lambda}\|y - v\|^2 + \frac{1}{2\mu}\|x - y\|^2\right]\right].$$

We consider the minimization problem
$$(3.28) \qquad m = \inf_{y \in X}\left[\frac{1}{2\lambda}\|y - v\|^2 + \frac{1}{2\mu}\|x - y\|^2\right]$$

Evidently the infimum is realized at $y \in X$ such that
$$\frac{1}{\lambda}\mathcal{F}(y - v) = \frac{1}{\mu}\mathcal{F}(x - y)$$
and, since $\mathcal{F}$ is positively homogeneous and one-to-one, this relation implies
$$y = \frac{\lambda x + \mu v}{\lambda + \mu}.$$

Using this in (3.28), we obtain

$$m = \frac{\lambda}{2(\lambda+\mu)^2}\|x-v\|^2 + \frac{\mu}{2(\lambda+\mu)^2}\|x-v\|^2$$
$$= \frac{1}{2(\lambda+\mu)}\|x-v\|^2.$$

So, it follows that

$$(\psi_\lambda)_\mu(x) = \inf_{v\in X}\left[\psi(v) + \frac{1}{2(\lambda+\mu)}\|x-v\|^2\right]$$
$$= \psi_{\lambda+\mu}(x), \ \forall x \in X,$$

that is, $(\psi_\lambda)_\mu = \psi_{\lambda+\mu}$. □

An immediate consequence of Lemma 6 is the following proposition:

PROPOSITION 7. *If $\psi \in \Gamma_0(X)$, then*

$$\varphi_\lambda \xrightarrow{M} \varphi \text{ as } \lambda \downarrow 0.$$

PROOF. From Lemma 6 we know that for every $\mu > 0$ and every $x \in X$, we have

$$(\varphi_\lambda)_\mu(x) = \varphi_{\lambda+\mu}(x) = (\varphi_\mu)_\lambda(x) \to \varphi_\mu(x) \text{ as } \lambda \downarrow 0,$$

which implies

$$\varphi_\lambda \xrightarrow{M} \varphi \text{ as } \lambda \downarrow 0,$$

(see Chapter 2). □

Now we have all the necessary tools to prove the generalization of Amann's theorem.

THEOREM 8. *If X is a reflexive Banach space which is embedded compactly and densely in $L^p(Z)$ $(1 < p < \infty)$ and $\widehat{\varphi} = \theta + J_0 : X \to \mathbb{R}$ is locally Lipschitz where $\theta \in C^1(X)$, $\nabla \theta = \widehat{A}$ is a bounded, $(S)_+ -$ operator, $J_0 = \widehat{J}_0|_X$ with $\widehat{J}_0$ as in (3.1), $\psi \in \Gamma_0(X)$, $\psi \geq 0$ and $x_0 \in X$ is an isolated minimizer of $\widehat{\varphi} + \psi$, then one can find $\widehat{r} > 0$ such that*

$$\deg(\partial\widehat{\varphi} + \partial\psi, B_{\widehat{r}}(x_0), 0) = 1$$

(here $\partial\widehat{\varphi}$ is the generalized subdifferential of $\widehat{\varphi}$ and $\partial\psi$ is the convex subdifferential of ψ).

PROOF. Clearly, without any loss of generality we may assume $(\widehat{\varphi}+\psi)(x_0) = 0$. Let $r_0 > 0$ be such that $x_0 \in X$ is the unique solution of $0 \in \partial\widehat{\varphi}(x) + \partial\psi(x)$ in $\overline{B}_{r_0}(x_0)$. Let $r_1 < r_0$ and set

$$\gamma = \inf\left\{\widehat{\varphi}(x) + \psi(x) : x \in \overline{B}_{r_0}(x_0) \setminus \overline{B}_{\frac{r_1}{2}}(x_0)\right\}$$

Evidently we have

$$\{\widehat{\varphi}+\psi < \gamma\} \subseteq B_{\frac{r_1}{2}}(x_0).$$

So we can find $\varepsilon > 0$ small such that

$$\{\widehat{\varphi}+\psi < \gamma\}_\varepsilon \subseteq B_{r_1}(x_0).$$

Recall that
$$\{\widehat{\varphi}+\psi<\gamma\}_{\varepsilon}:=\{x\in X:d\left(x,\{x:\widehat{\varphi}\left(x\right)+\psi\left(x\right)<\gamma\}\right)<\varepsilon\}.$$

We let $\mu<\gamma$ and $r_1<r_2<r_0$. From Proposition 7 we know that
$$\widehat{\varphi}+\psi_{\lambda}\xrightarrow{M}\widehat{\varphi}+\psi\text{ as }\lambda\downarrow 0.$$

Then from Mosco [**39**] (see also Denkowski-Migorski-Papageorgiou [**12**], p.468) we have that
$$\{\widehat{\varphi}+\psi_{\lambda}\leq\mu\}\xrightarrow{M}\{\widehat{\varphi}+\psi\leq\mu\}\text{ as }\lambda\downarrow 0.$$

It follows that

(3.29) $\quad w-\limsup\limits_{\lambda\downarrow 0}\;\{\widehat{\varphi}+\psi_{\lambda}\leq\mu\}\cap\overline{B}_{r_2}\left(x_0\right)\subseteq\{\widehat{\varphi}+\psi_{\lambda}\leq\mu\}\cap\overline{B}_{r_2}\left(x_0\right).$

Note that
$$\{\widehat{\varphi}+\psi_{\lambda}\leq\mu\}\cap\overline{B}_{r_2}\left(x_0\right)\subseteq\{\widehat{\varphi}+\psi<\gamma\}\cap\overline{B}_{r_2}\left(x_0\right)$$
$$\subseteq\{\widehat{\varphi}+\psi<\gamma\}_{\varepsilon}\subseteq B_{r_1}\left(x_0\right).$$

From this inclusion and (3.29), we see that for $\lambda>0$ small, we have

(3.30) $\qquad\{\widehat{\varphi}+\psi_{\lambda}\leq\mu\}\cap\overline{B}_{r_2}\left(x_0\right)\subseteq\{\widehat{\varphi}+\psi<\gamma\}_{\varepsilon}\subseteq B_{r_1}\left(x_0\right).$

Next let us choose $r\in\left(0,\frac{r_1}{2}\right)$ and $\xi>0$ such that
$$\xi<\frac{1}{2}\inf\left\{\widehat{\varphi}\left(x\right)+\psi\left(x\right):x\in B_{r_0}\left(x_0\right)\backslash B_r\left(x_0\right)\right\}\leq\frac{\mu}{2}.$$

Evidently, we have
$$\{\widehat{\varphi}+\psi\leq\xi\}\subseteq B_r\left(x_0\right)\subseteq\overline{B}_r\left(x_0\right)\subseteq B_{r'}\left(x_0\right)\text{ with }r'<\frac{r_1}{2}.$$

Then for $\varepsilon>0$ small we have
$$\{\widehat{\varphi}+\psi\leq\xi\}_{\varepsilon}\subseteq B_{r'}\left(x_0\right).$$

From this inclusion and recalling that
$$\{\widehat{\varphi}+\psi_{\lambda}\leq\xi\}\xrightarrow{M}\{\widehat{\varphi}+\psi\leq\xi\}\text{ as }\lambda\downarrow 0,$$

for $\lambda>0$ small, we conclude that

(3.31) $\qquad\qquad\qquad\{\widehat{\varphi}+\psi_{\lambda}\leq\xi\}\subseteq B_{r'}\left(x_0\right).$

We set
$$\widehat{m}=\sup\;\{\|x^*\|_*:x^*\in\partial\widehat{\varphi}\left(B_{r_0}\left(x_0\right)\right)\}$$
and consider $x\in\{\widehat{\varphi}+\psi_{\lambda}\leq\xi\}$. For $t\in[0,1]$, we let
$$x_t=tx_0+(1-t)\,x.$$

We will show that $x_t\in V_{\lambda}=\{\widehat{\varphi}+\psi_{\lambda}<\mu\}$ for some $\lambda>0$ small. To this end, from the nonsmooth mean value theorem, we know that given $y,z\in B_{r_0}\left(x_0\right)$, we can find $\beta\in(0,1)$ and $u^*\in\partial\widehat{\varphi}(\beta y+(1-\beta)\,z)$ such that
$$\widehat{\varphi}\left(y\right)-\widehat{\varphi}\left(z\right)=\left(u^*,y-z\right).$$

Then

(3.32) $\quad\widehat{\varphi}\left(z\right)-\widehat{m}\left\|y-z\right\|\leq\widehat{\varphi}\left(y\right)\leq\widehat{\varphi}\left(z\right)+\widehat{m}\left\|y-z\right\|\;\forall y,z\in B_{r_0}\left(x_0\right).$

We have

$$\widehat{\varphi}(tx_0 + (1-t)x)$$
$$= t\widehat{\varphi}(tx_0 + (1-t)x) + (1-t)\widehat{\varphi}(x + t(x_0 - x))$$
$$\leq t\widehat{\varphi}(x_0) + t(1-t)\widehat{m}\|x - x_0\| + (1-t)\widehat{\varphi}(x) + t(1-t)\widehat{m}\|x - x_0\|$$
(3.33) $$= t\widehat{\varphi}(x_0) + (1-t)\widehat{\varphi}(x) + 2t(1-t)\widehat{m}\|x - x_0\|.$$

The Moreau-Yosida regularization ψ_λ is convex. So

(3.34) $$\psi_\lambda(x_t) \leq t\psi_\lambda(x_0) + (1-t)\psi_\lambda(x).$$

Adding (3.33) and (3.34), we obtain

(3.35) $$(\widehat{\varphi} + \psi_\lambda)(x_t) \leq \xi + 2t(1-t)\widehat{m}\|x - x_0\|$$

Fix $0 < r < r' < \min\left\{\frac{r_1}{2}, \frac{\mu}{\widehat{m}}\right\}$. Then from (3.35) and for $\lambda > 0$ small, we have

$$(\widehat{\varphi} + \psi_\lambda)(x_t) \leq \xi + 2t(1-t)\widehat{m}r' \leq \xi + \frac{1}{2}\widehat{m}\frac{\mu}{2\widehat{m}} < \frac{\mu}{2} + \frac{\mu}{2} = \mu,$$

hence

(3.36) $$x_t \in V_\lambda = \{\widehat{\varphi} + \psi_\lambda < \mu\} \text{ for } \lambda > 0 \text{ small}.$$

Finally we show that

(3.37) $$0 \notin \partial\widehat{\varphi}(x) + \partial\psi_\lambda(x) \ \forall x \in \{\xi \leq \widehat{\varphi} + \psi_\lambda\} \cap \overline{B}_{r_2}(x_0), \ \forall \lambda > 0 \text{ small}.$$

We argue indirectly. Suppose that (3.37) is not true. Then we can find $\lambda_n \downarrow 0$ and $x_n \in \{\xi \leq \widehat{\varphi} + \psi_{\lambda_n}\} \cap \overline{B}_{r_2}(x_0)$ such that

$$0 \in \partial\widehat{\varphi}(x_n) + \partial\psi_{\lambda_n}(x_n) = \partial\widehat{\varphi}(x_n) + \nabla\psi_{\lambda_n}(x_n),$$

hence

(3.38) $$\widehat{A}(x_n) + u_n^* = -\nabla\psi_{\lambda_n}(x_n)$$

where $u_n^* \in L^q(Z)$ $\left(\frac{1}{p} + \frac{1}{q} = 1\right)$ and $u_n^*(z) \in \partial j_0(z, x_n(z))$ a.e. on Z for all $n \geq 1$. Clearly $\{x_n\}_{n\geq 1}$ is bounded in X and so it follows that $\{\nabla\psi_{\lambda_n}(x_n)\}_{n\geq 1} \subseteq X^*$ is bounded too. By passing to a suitable subsequence if necessary, we may assume that

$$x_n \xrightarrow{w} x \text{ in } X \text{ and } \nabla\psi_{\lambda_n}(x_n) \xrightarrow{w} v^* \text{ in } X^*.$$

Since X is embedded compactly into $L^p(Z)$, we obtain

$$x_n \longrightarrow x \text{ in } L^p(Z).$$

We take duality brackets of (3.38) with $x_n - x$ and so we obtain

$$\left\langle \widehat{A}(x_n), x_n - x \right\rangle + \int_Z u_n^*(z)(x_n - x)(z)\,dz$$
(3.39) $$= -\left\langle \nabla\psi_{\lambda_n}(x_n), x_n - x \right\rangle \text{ for all } n \geq 1.$$

Since ψ_{λ_n} is a convex, Gâteaux differentiable function, we have

(3.40) $$-\left\langle \nabla\psi_{\lambda_n}(x_n), x_n - x \right\rangle \leq \psi_{\lambda_n}(x) - \psi_{\lambda_n}(x_n).$$

From Proposition 7 we know that $\psi_{\lambda_n} \xrightarrow{M} \psi$. Since $x_n \xrightarrow{w} x$ in X, we have

(3.41) $$\psi(x) \leq \liminf_{n\to\infty} \psi_{\lambda_n}(x_n)$$

(see Chapter 2). Passing to the limit as $n \to \infty$ in (3.40) and using (3.41) and the fact that $\psi_{\lambda_n}(x_n) \leq \psi(x)$ for all $n \geq 1$, we obtain

(3.42) $$\limsup_{n \to \infty} -\langle \nabla \psi_{\lambda_n}(x_n), x_n - x \rangle \leq \psi(x) - \psi(x) = 0.$$

Since $x_n \longrightarrow x$ in $L^p(Z)$ and $\{u_n^*\}_{n \geq 1} \in L^q(Z)$ is bounded (see hypothesis $(H_{j_0})(iii)$) we infer that

(3.43) $$\int_Z u_n^*(z)(x_n(z) - x(z))\, dz \to 0 \text{ as } n \to \infty.$$

So if we pass to the limit as $n \to \infty$ in (3.39) and we use (3.42) and (3.43), we obtain
$$\limsup_{n \to \infty} \langle \widehat{A}(x_n), x_n - x \rangle \leq 0.$$

By hypothesis, $\widehat{A}$ is of type $(S)_+$. So, it follows that
$$x_n \longrightarrow x \text{ in } X.$$

Recall that $\psi_{\lambda_n} \xrightarrow{M} \psi$ (see Proposition 7). Therefore it follows that
$$\partial \psi_{\lambda_n} = \{\nabla \psi_{\lambda_n}\} \xrightarrow{K_{ws}} \partial \psi$$

(see Chapter 2). Since $x_n \longrightarrow x$ in X and $\nabla \psi_{\lambda_n} \xrightarrow{w^*} v^*$ in X^*, we infer that
$$x^* \in \partial \psi(x).$$

By virtue of the convexity of ψ_{λ_n} we have

(3.44) $$\langle \nabla \psi_{\lambda_n}(x_n), x - x_n \rangle \leq \psi_{\lambda_n}(x) - \psi_{\lambda_n}(x_n), \ n \geq 1.$$

Note that $\langle \nabla \psi_{\lambda_n}(x_n), x - x_n \rangle \to 0$ and $\psi_{\lambda_n}(x) \to \psi(x)$ as $n \to \infty$. So, from (3.44), we obtain
$$\limsup_{n \to \infty} \psi_{\lambda_n}(x_n) \leq \psi(x).$$

On the other hand, since $\psi_{\lambda_n} \xrightarrow{M} \psi$ (see Proposition 7), we have
$$\psi(x) \leq \liminf_{n \to \infty} \psi_{\lambda_n}(x_n).$$

Therefore, we see that
$$\psi_{\lambda_n}(x_n) \to \psi(x).$$

So, in the limit as $n \to \infty$, we have
$$0 < \xi \leq \widehat{\varphi}(x) + \psi(x)$$

hence $x \neq x_0$ (since by hypothesis $\widehat{\varphi}(x_0) + \psi(x_0) = 0$) and $x \in \overline{B}_{r_2}(x_0)$. Since $\{u_n^*\}_{n \geq 1} \subseteq L^q(Z)$ is bounded, we may assume that
$$u_n^* \xrightarrow{w} u^* \text{ in } L^q(Z).$$

We know that
$$u_n^*(z) \in \partial j_0(z, x_n(z)) \text{ a.e. on } Z \text{ and } x_n \longrightarrow x \text{ in } X.$$

Exploiting the upper semicontinuity of $\partial j_0(z, \cdot)$ and using Mazur's lemma we obtain
$$u^*(z) \in \partial j_0(z, x(z)) \text{ a.e. on } Z.$$

If we pass to the limit in (3.38) as $n \to \infty$, we get
$$\widehat{A}(x) + u^* + v^* = 0.$$

But $\widehat{A}(x) + u^* \in \partial\widehat{\varphi}(x)$ and $v^* \in \partial\psi(x)$. Hence
$$0 \in \partial\widehat{\varphi}(x) + \partial\psi(x) \text{ with } x \neq x_0, \, x \in \overline{B}_{r_2}(x_0),$$
a contradiction to the fact that in $\overline{B}_{r_0}(x_0)$ only x_0 satisfies this inclusion. We know that $\partial\psi$ is maximal monotone and for every $\lambda > 0$
$$(\partial\psi)_\lambda = \partial\psi_\lambda = \nabla\psi_\lambda.$$
Since $\nabla\psi_\lambda$ is everywhere defined, bounded, continuous and maximal monotone, we can easily check that $\widehat{A} + \nabla\psi_\lambda$ is bounded continuous and of type $(S)_+$. Also let
$$N_0 = \partial J_0 = \partial \widehat{J}_0.$$
We know that N_0 belongs to class (P) (see Corollary 4). Note that
$$\partial\widehat{\varphi} = A + N_0.$$
Let $\lambda > 0$ be small so that (3.30), (3.31) and (3.37) hold. Also let $U = B_{r_2}(x_0)$ and let $\widehat{g}_\delta : X \to X^*$ be a continuous δ-approximate selection of N_0. Then for $\delta > 0$ small we have
$$\deg(\partial\widehat{\varphi} + \nabla\psi_\lambda, U, 0) = \deg\left(\widehat{A} + \nabla\psi_\lambda + N_0, U, 0\right)$$
(3.45)
$$= \deg\left(\widehat{A} + \nabla\psi_\lambda + g_\delta, U, 0\right)$$
(see Chapter 2). From the excision property of the degree map we have
(3.46) $\qquad \deg(\partial\widehat{\varphi} + \nabla\psi_\lambda, U, 0) = \deg(\partial\widehat{\varphi} + \nabla\psi_\lambda, V_\lambda, 0).$

If we set $\widehat{\varphi}^\lambda = \theta^\lambda + J_0$ with $\theta^\lambda = \theta + \psi_\lambda$, then we satisfy the requirements of Proposition 5 with $\widehat{\varphi} = \widehat{\varphi}^\lambda$ and $V = V_\lambda$ and so
(3.47) $\qquad \deg(\partial\widehat{\varphi} + \nabla\psi_\lambda, V_\lambda, 0) = 1.$

From (3.45), (3.46) and (3.47), it follows that for $\lambda > 0$ and $\delta > 0$ small, we have
$$\deg(\partial\widehat{\varphi} + \nabla\psi_\lambda, U, 0) = 1,$$
hence
$$\deg(\partial\widehat{\varphi} + \partial\psi, U, 0) = 1.$$
So if we set $\widehat{r} = r_2$, we have proved the theorem. $\square$

CHAPTER 4

Variational-Hemivariational Inequalities

In this Chapter, we use Theorem 8 to prove multiplicity results for some variational inequalities with a nonsmooth potential (variational-hemivariational inequalities). As before, let $Z \subseteq \mathbb{R}^N$ be a bounded domain with a C^2–boundary ∂Z.

First we consider the following obstacle problem

(4.1) $\begin{cases} \int_Z \|Dx(z)\|_{\mathbb{R}^N}^{p-2} \left(Dx(z), Dy(z) - Dx(z)\right)_{\mathbb{R}^N} dz \\ \qquad \geq \int_Z u(z)(y(z) - x(z)) dz \text{ for all } y \in C, \\ 1 < p < \infty, \, u \in L^q(Z), \, (\frac{1}{p} + \frac{1}{q} = 1), \\ u(z) \in \partial j(z, x(z)) \text{ a.e. on } Z \text{ and} \\ C = \left\{ y \in W_0^{1,p}(Z) : y(z) \geq \gamma(z) \text{ a.e. on } Z \right\}. \end{cases}$

The hypotheses on the data of (4.1) are the following:

(H_γ) $\gamma \in L^\infty(Z) \cap W_0^{1,p}(Z)$ and $\operatorname*{ess\,sup}_Z \gamma < 0$.

$(H_j)^1$ $j : Z \times \mathbb{R}$ is a function such that that $j(z, 0) = 0$ a.e. on Z and
  (i) for all $x \in \mathbb{R}$, $z \to j(z, x)$ is measurable;
  (ii) for almost all $z \in Z$, $x \to j(z, x)$ is locally Lipschitz;
  (iii) for almost all $z \in Z$, all $x \in \mathbb{R}$ and all $u \in \partial j(z, x)$, we have

  $|u| \leq a(z) + c|x|^{p-1}$, with $a \in L^\infty(Z)_+$ $c > 0$;

  (iv) there exists $\theta \in L^\infty(Z)_+$ such that $\theta(z) \leq \lambda_1$ a.e. on Z, with strict inequality on a set of positive measure ($\lambda_1 > 0$ is the principal eigenvalue of $\left(-\triangle_p, W_0^{1,p}(Z)\right)$ with weight $m = 1$) and

  $0 \leq \liminf_{x \to +\infty} \dfrac{u}{x^{p-1}} \leq \limsup_{x \to +\infty} \dfrac{u}{x^{p-1}} \leq \theta(z)$

  uniformly for almost all $z \in Z$ and all $u \in \partial j(z, x)$.

  (v) there exist $\eta, \widehat{\eta} \in L^\infty(Z)_+$ such that $\lambda_1 \leq \eta(z)$ a.e. on Z, with strict inequality on a set of positive measure, $\widehat{\eta}(z) < \lambda_2$ a.e. on Z and

  $\eta(z) \leq \liminf_{x \to 0} \dfrac{u}{|x|^{p-2}x} \leq \limsup_{x \to 0} \dfrac{u}{|x|^{p-2}x} \leq \widehat{\eta}(z)$

  uniformly for almost all $z \in Z$ and all $u \in \partial j(z, x)$ (where λ_2 denotes the second eigenvalue of $\left(-\triangle_p, W_0^{1,p}(Z)\right)$ with weight $m = 1$).

REMARK 9. *The hypotheses $(H_j)^1$ (iv) and (v) are nonuniform nonresonance conditions at $+\infty$ and at 0. The condition at $+\infty$ is from below $\lambda_1 > 0$ (the principal* eigenvalue of $\left(-\triangle_p, W_0^{1,p}(Z)\right)$) and the condition at 0 is from above $\lambda_1 > 0$. Also at 0 we have a uniform nonresonance from below $\lambda_2 > 0$.

REMARK 10. *A simple nonsmooth locally Lipschitz potential satisfying hypotheses* $(H_j)^1$ *is the following. For simplicity we drop the* $z-$ *dependence:*

$$j(x) = \begin{cases} \frac{\eta}{p}|x|^p - \frac{1}{p}\cos|x|^p & \text{if } |x| \leq 1 \\ \frac{\theta}{p}|x|^p + \frac{\eta-\theta}{p} - \frac{1}{p}\cos 1 & \text{if } |x| > 1 \end{cases}$$

with $\theta < \lambda_1 < \eta < \lambda_2$.

The Euler functional $\varphi : W_0^{1,p}(Z) \to \overline{\mathbb{R}} = \mathbb{R} \cup \{+\infty\}$ for problem (4.1) is defined by

$$\varphi(x) = \widehat{\varphi}(x) + \psi(x), \ x \in W_0^{1,p}(Z),$$

with

$$\widehat{\varphi}(x) = \frac{1}{p}\|Dx\|_p^p - \int_Z j(z, x(z)) \, dz, \ x \in W_0^{1,p}(Z),$$

and

$$\psi(x) = i_C(x) = \begin{cases} 0 & \text{if } x \in C \\ +\infty & \text{if } x \notin C \end{cases}.$$

Because of hypotheses $(H_j)^1$ (i), (ii) (iii), $\widehat{\varphi}$ is Lipschitz continuous on bounded sets, hence locally Lipschitz (see Gasinski-Papageorgiou [**18**], p.59). Also $C \subseteq W_0^{1,p}(Z)$ is closed, convex, hence $\psi \in \Gamma_0\left(W_0^{1,p}(Z)\right)$.

Throughout this Chapter by $\langle .,. \rangle$ we denote the duality brackets for the pair $\left(W^{-1,q}(Z), W_0^{1,p}(Z)\right)$.

We consider the nonlinear operator $A : W_0^{1,p}(Z) \to W^{-1,q}(Z)$ defined by

(4.2) $\quad \langle A(x), y \rangle = \int_Z \|Dx(z)\|_{\mathbb{R}^N}^{p-2} (Dx(z), Dy(z))_{\mathbb{R}^N} \, dz$ for $x, y \in W_0^{1,p}(Z)$.

LEMMA 11. $A : W_0^{1,p}(Z) \to W^{-1,q}(Z)$ *is a maximal monotone, bounded and continuous operator of type* $(S)_+$.

PROOF. It is clear from (4.2) that the operator A is bounded, continuous and monotone, hence it is maximal monotone (see Gasinski-Papageorgiou [**18**], p.75). Note that A is the Fréchet derivative of $x \to \frac{1}{p}\|Dx\|_p^p$, viewed as a functional on $W_0^p(Z)$. Next suppose that $x_n \xrightarrow{w} x$ in $W_0^{1,p}(Z)$ and that

$$\limsup_{n \to \infty} \langle A(x_n), x_n - x \rangle \leq 0.$$

We need to show that $x_n \to x$ in $W_0^{1,p}(Z)$. Since A is maximal monotone, it is generalized pseudomonotone (see Gasinski-Papageorgiou [**18**], p.84). Note that $\{A(x_n)\}_{n \geq 1}$ is bounded and so by the Eberlein-Smulian theorem, we can find a subsequence $\{x_{n_k}\}_{k \geq 1}$ of $\{x_n\}_{n \geq 1}$ such that

$$A(x_{n_k}) \xrightarrow{w} v^* \text{ in } W^{-1,q}(Z).$$

Because A is generalized pseudomonotone, we have that

$$v^* = A(x) \text{ and } \langle A(x_{n_k}), x_{n_k} \rangle \to \langle A(x), x \rangle.$$

So we have
$$\|Dx_{n_k}\|_p \to \|Dx\|_p.$$

Since $Dx_{n_k} \xrightarrow{w} Dx$ in $L^p(Z, \mathbb{R}^N)$ and $L^p(Z, \mathbb{R}^N)$ being uniformly convex, it has the Kadec-Klee property, we infer that $Dx_{n_k} \to Dx$ in $L^p(Z, \mathbb{R}^N)$, hence $x_{n_k} \to x$ in $W_0^{1,p}(Z)$. We may therefore conclude that $x_n \to x$ in $W_0^{1,p}(Z)$. □

The next Lemma highlights the hypothesis $(H_j)^1 (iv)$.

LEMMA 12. *If $\theta \in L^\infty(Z)_+$ with $\theta(z) \leq \lambda_1$ a.e. on Z and the inequality is strict on a set of positive measure, then one can find $\xi > 0$ such that*
$$\tau(x) = \|Dx\|_p^p - \int_Z \theta(z)|x(z)|^p\,dz \geq \xi \|Dx\|_p^p \text{ for all } x \in W_0^{1,p}(Z).$$

PROOF. Because of (2.1) and the hypotheses on θ, we have that $\tau \geq 0$. Suppose that the lemma is not true. Then exploiting the p-homogeneity of τ, we can find $\{x_n\}_{n \geq 1} \subseteq W_0^{1,p}(Z)$ such that
$$\|Dx_n\|_p = 1 \text{ for all } n \geq 1, \text{ and } \tau(x_n) \downarrow 0.$$

By virtue of Poincaré's inequality, $\{x_n\}_{n \geq 1} \subseteq W_0^{1,p}(Z)$ is bounded and so we may assume that
$$x_n \xrightarrow{w} x \text{ in } W_0^{1,p}(Z),$$
$$x_n \to x \text{ in } L^p(Z),$$
$$x_n(z) \to x(z) \text{ a.e. on } Z$$
and
$$|x_n(z)| \leq k(z) \text{ a.e. on } Z \text{ with } k \in L^p(Z)$$
(recall that $W_0^{1,p}(Z)$ is embedded compactly in $L^p(Z)$ and use Proposition 2.2.41, p.147, of Denkowski-Migorski-Papageorgiou [13]). Since the norm in a Banach space is weakly lower semicontinuous we have
$$\|Dx\|_p^p \leq \liminf_{n \to \infty} \|Dx_n\|_p^p.$$

Also from the Lebesgue dominated convergence theorem, we have
$$\int_Z \theta(z)|x_n(z)|^p\,dz \to \int_Z \theta(z)|x(z)|^p\,dz.$$

So finally
$$\tau(x) \leq \lim_{n \to \infty} \tau(x_n)$$
hence
(4.3) $$\|Dx\|_p^p \leq \int_Z \theta(z)|x_n(z)|^p\,dz \leq \lambda_1 \|x\|_p^p.$$

Because of (2.1) (with $m \equiv 1$), we see that
$$\|Dx\|_p^p = \lambda_1 \|x\|_p^p$$
hence
$$x = 0 \text{ or } x = \pm u_1.$$

If $x = 0$, then $\|Dx_n\|_p \to 0$, a contradiction to the fact that $\|Dx_n\|_p = 1$. So $x = \pm u_1$ and $|x(z)| = |u_1(z)| > 0$ for all $z \in Z$. Then from the first inequality in (4.3) and the hypotheses on θ, we obtain
$$\|Dx\|_p^p < \lambda_1 \|x\|_p^p$$
which contradicts (2.1) (with $m \equiv 1$). Therefore the Lemma is proved. $\square$

PROPOSITION 13. *If hypotheses* (H_γ) *and* $(H_j)^1 (i)-(iv)$ *hold, then there exists* $x_0 \in C$ *such that*
$$\varphi(x_0) = \inf_{x \in W_0^{1,p}(Z)} \varphi(x).$$

PROOF. By virtue of hypothesis $(H_j)^1 (iv)$, given $\varepsilon > 0$, we can find $M_1 = M_1(\varepsilon) > 0$ such that for almost all $z \in Z$, all $x \geq M_1$ and all $u \in \partial j(z,x)$, we have
$$(4.4) \qquad u \leq (\theta(z) + \varepsilon) x^{p-1}.$$
Also, because of hypothesis $(H_j)^1 (iii)$, we can find $\beta_\varepsilon \in L^\infty(Z)_+$ such that for almost all $z \in Z$, all $x \in [-\|\gamma\|_\infty, M_1]$ and all $u \in \partial j(z,x)$, we have
$$(4.5) \qquad |u| \leq \beta_\varepsilon(z).$$
By Rademacher's theorem for a.a. $z \in Z$, $j(z,.)$ is differentiable almost everywhere and
$$\frac{d}{dr} j(z,r) \in \partial j(z,r).$$
Therefore, for almost all $z \in Z$ and for all $x \geq 0$, we have
$$j(z,x) = \int_0^x \frac{d}{dr} j(z,r) \, dr$$
$$\leq \int_0^x (\theta(z) + \varepsilon) r^p \, dr + \beta_\varepsilon(z) x \quad (\text{see } (4.4) \text{ and } (4.5))$$
$$(4.6) \qquad = \frac{1}{p} (\theta(z) + \varepsilon) x^p + \beta_\varepsilon(z) x.$$
Also if $x \in [-\|\gamma\|_\infty, 0]$, then
$$(4.7) \qquad -j(z,x) = \int_x^0 \frac{d}{dr} j(z,r) \geq -\beta_\varepsilon(z) x \quad (\text{see } (4.5)).$$
For every $x \in C$, by (4.6) and (4.7) we have
$$\varphi(x) = \frac{1}{p} \|Dx\|_p^p - \int_Z j(z, x(z)) \, dz$$
$$= \frac{1}{p} \|Dx\|_p^p - \int_{\{x \geq 0\}} j(z, x(z)) \, dz - \int_{\{-\|\gamma\|_\infty \leq x < 0\}} j(z, x(z)) \, dz$$
$$\geq \frac{1}{p} \|Dx\|_p^p - \frac{1}{p} \int_Z \theta(z) |x(z)|^p \, dz - \frac{\varepsilon}{p} \|x\|_p^p - c_1 \|Dx\|_p - c_2$$
$$= \frac{1}{p} \left(\xi - \frac{\varepsilon}{\lambda_1} \right) \|Dx\|_p^p - c_1 \|Dx\|_p - c_2$$
for some $c_1, c_2 > 0$, (see Lemma 12 and (2.1)). So, if we choose $\varepsilon < \lambda_1 \xi$ it follows that φ is coercive. Also, exploiting the compact embeeding of $W_0^{1,p}(Z)$ into $L^p(Z)$, it is easy to check that φ is weakly lower semicontinuous on $W_0^{1,p}(Z)$.

Therefore, by the Weierstrass theorem, we can find $x_0 \in W_0^{1,p}(Z)$ such that $\varphi(x_0) = \inf_{x \in W_0^{1,p}(Z)} \varphi(x)$. □

In the next Proposition we will use hypothesis $(H_j)^1(v)$ to show that for small balls the degree map of $\partial \widehat{\varphi} + \partial \psi$ is -1.

PROPOSITION 14. *If hypotheses (H_γ) and $(H_j)^1$ hold, then there exists $\rho_0 > 0$ such that for all $0 < \rho \leq \rho_0$, we have*
$$\deg(\partial \widehat{\varphi} + \partial \psi, B_\rho(0), 0) = -1.$$

PROOF. Let $h \in L^\infty(Z)_+$ be such that $\eta(z) \leq h(z) \leq \widehat{\eta}(z)$ a.e. on Z. Also for every $t \in [0, 1]$ we set
$$C(t) = \left\{ y \in W_0^{1,p}(Z) : y(z) \geq \frac{\gamma(z)}{t} \text{ a.e. on } Z \right\}, \; t \in (0, 1],$$
$$C(0) = W_0^{1,p}(Z),$$
and, for all $x \in W_0^{1,p}(Z)$,
$$\psi_t(x) = i_{C(t)}(x) = \begin{cases} 0 & \text{if } x \in C(t) \\ +\infty & \text{if } x \notin C(t) \end{cases}.$$

Evidently $C(1) = C$. If $t_n \to t$ and $x \in w - \limsup_{n \to \infty} C(t_n)$, then we can find a subsequence $\{t_{n_k}\}_{k \geq 1}$ of $\{t_n\}_{n \geq 1}$ and $x_{n_k} \in C(t_{n_k})$ such that $x_{n_k} \xrightarrow{w} x$ in $W_0^{1,p}(Z)$. Since $x_{n_k} \to x$ in $L^p(Z)$ (by the compact embedding of $W_0^{1,p}(Z)$ in $L^p(Z)$), we may assume that
$$x_{n_k}(z) \to x(z) \text{ a.e. on } Z.$$
We have that
$$x_{n_k}(z) \geq \frac{\gamma(z)}{t_{n_k}} \text{ a.e. on } Z.$$
and in the limit we obtain
$$x(z) \geq \frac{\gamma(z)}{t} \text{ a.e. on } Z \text{ if } t \neq 0$$
and
$$x(z) \geq -\infty \text{ a.e. on } Z \text{ if } t = 0$$
(recall that $\gamma(z) < 0$ a.e. on Z). Therefore $x \in C(t)$. Next, let $x \in C(t), t \in (0, 1]$, $t_n \to t$ and
$$x_n(z) = x(z) + \left(\frac{1}{t_n} - \frac{1}{t} \right) \gamma(z) \geq \frac{\gamma(z)}{t_n} \text{ a.e. on } Z$$
i.e., $x_n \in C(t_n)$ for all $n \geq 1$. Moreover, we have
$$\|x_n - x\|_{W_0^{1,p}(Z)} \leq \frac{|t - t_n|}{t t_n} \|\gamma\|_{W_0^{1,p}(Z)} \to 0 \text{ as } n \to \infty.$$
So finally we conclude that
$$C(t_n) \xrightarrow{M} C(t) \text{ in } W_0^{1,p}(Z),$$
hence
$$\psi_{t_n} \xrightarrow{M} \psi(t)$$

(see Chapter 2). Let $x \in C(0) = W_0^{1,p}(Z)$. Then we can find $\{x_k\}_{k\geq 1} \subseteq C_c^1(Z)$ such that
$$x_k \to x \text{ in } W_0^{1,p}(Z).$$
Since by hypothesis $\gamma \in -int\ L^\infty(Z)_+$, given any $y \in C_c^1(Z)$ we can find $\mu(y) > 0$ such that
$$\mu(y)\gamma - y \in -int L^\infty(Z)_+.$$
Note that $\frac{1}{t_n} \to +\infty$ (since $t_n \to 0^+$). So for every $k \geq 1$ we can find $n_k > 1$ large such that
$$\frac{1}{t_{n_k}}\gamma \leq x_k$$
hence
$$x_k \in C(t_{n_k}) \text{ and } x_k \to x \text{ in } W_0^{1,p}(Z).$$
This together with the earlier part of the proof proves that
$$C(t_{n_k}) \xrightarrow{M} C(0).$$
Then every subsequence of $\{C(t_n)\}_{n\geq 1}$ has a further subsequence converging in the Mosco sense to $C(0)$. But the Mosco convergence of nonempty, closed and convex sets in $W_0^{1,p}(Z)$ corresponds to a Polish topology. So by Urysohn's criterion for convergent sequences we conclude that
$$C(t_n) \xrightarrow{M} C(0) \text{ if } t_n \to 0.$$
Therefore $\{\partial \psi_t\}_{t \in [0,1]}$ is a pseudomonotone homotopy (see Kobayashi-Otani [27], Theorem 3.5). Also let $K: W_0^{1,p}(Z) \to W^{-1,q}(Z)$ be defined by
$$K(x)(.) = |x(.)|^{p-2} x(.).$$
Since $K\left(W_0^{1,p}(Z)\right) = L^q(Z)$ ($\frac{1}{p} + \frac{1}{q} = 1$) and $L^q(Z)$ is embedded compactly in $W^{-1,q}(Z)$, it follows that K is completely continuous. Consider the homotopy $\widehat{h}: [0,1] \times W_0^{1,p}(Z) \to 2^{W^{-1,q}(Z)}\setminus \{\varnothing\}$ defined by
$$\widehat{h}(t,x) = A(x) - tN(x) - (1-t)hK(x)$$
where $N: W_0^{1,p}(Z) \to 2^{W^{-1,q}(Z)} \setminus \{\varnothing\}$ is defined by
$$N(x) = \{u \in L^q(Z) : u(z) \in \partial j(z, x(z)) \text{ a.e. on } Z\}.$$
From Corollary 4, Lemma 11 and the complete continuity of K, we check easily that $\widehat{h}$ is an $(S)_+$-homotopy. Let
$$\widetilde{h}(t,x) = \widehat{h}(t,x) + \partial \psi_t(x) \text{ for } (t,x) \in [0,1] \times W_0^{1,p}(Z).$$
Claim: There exists $\rho_0 > 0$ such that for all $t \in [0,1]$, all $0 < \rho \leq \rho_0$ and all $x \in \partial B_\rho(0) \subseteq W_0^{1,p}(Z)$ we have
$$0 \notin \widetilde{h}(t,x).$$
Suppose that the Claim is not true. Then we can find $\{t_n\}_{n\geq 1} \subseteq [0,1]$ and $x_n \in C(t_n)$, $n \geq 1$, such that
$$t_n \to t \text{ in } [0,1], \ \|x_n\| \to 0$$
and
(4.8) $\quad 0 \in A(x_n) - t_n N(x_n) - (1 - t_n)hK(x_n) + \partial \psi_{t_n}(x_n), \ n \geq 1.$

Here and in the sequel $\|x_n\| = \|x_n\|_{W_0^{1,p}(Z)}$. Let
$$v_n = \frac{x_n}{\|x_n\|}, \ n \geq 1.$$
By passing to a suitable subsequence if necessary, we may assume that
$$v_n \xrightarrow{w} v \text{ in } W_0^{1,p}(Z) \text{ and } v_n \to v \text{ in } L^p(Z).$$
Note that if $y \in C(t_n)$, then
$$\widehat{y}(z) := \frac{y(z)}{\|x_n\|} \geq \frac{1}{t_n \|x_n\|} \gamma(z) \text{ a.e. on } Z,$$
i.e., $\widehat{y} \in C(t_n \|x_n\|)$. Conversely, if $\widehat{y} \in C(t_n \|x_n\|)$, then $y = \|x_n\| \widehat{y} \in C(t_n)$ for all $n \geq 1$, therefore
$$\|x_n\| C(t_n \|x_n\|) = C(t_n) \text{ for all } n \geq 1.$$
From (4.8), we have
$$-A(x_n) + t_n u_n + (1 - t_n) hK(x_n) \in \partial \psi_{t_n}(x_n), \text{ with } u_n \in N(x_n),$$
hence,
$$\langle A(x_n), y - x_n \rangle - t_n \int_Z u_n (y - x_n) dz - (1 - t_n) \int_Z h |x_n|^{p-2} x_n (y - x_n) dz \geq 0.$$
for all $y \in C(t_n)$. Dividing this inequality with $\|x_n\|^p$, we obtain

(4.9)
$$\begin{array}{c} \langle A(v_n), \widehat{y} - v_n \rangle - t_n \int_Z \frac{u_n}{\|x_n\|^{p-1}} (\widehat{y} - v_n) dz \\ - (1 - t_n) \int_Z h |v_n|^{p-2} v_n (\widehat{y} - v_n) dz \geq 0 \end{array}$$

for all $\widehat{y} \in C(t_n \|x_n\|)$. Because of hypothesis $(H_j)^1(v)$, we can find a $\delta > 0$ such that for almost all $z \in Z$, all x with $|x| < \delta$ and all $u \in \partial j(z, x)$, we have

(4.10) $$|u| \leq (\widehat{\eta}(z) + 1) |x|^{p-1}.$$

On the other hand, by assumption $(H_j)^1(iii)$, for almost all $z \in Z$, and all $x \in \mathbb{R}$ with $|x| \geq \delta$ and all $u \in \partial j(z, x)$ we have

(4.11) $$|u| \leq a(z) + c |x|^{p-1} \leq \left(\frac{a(z)}{\delta^{p-1}} + c \right) |x|^{p-1}.$$

From (4.10) and (4.11) it follows that for almost all $z \in Z$, all $x \in \mathbb{R}$ and all $u \in \partial j(z, x)$, we have

(4.12) $$|u| \leq c_1 |x|^{p-1} \text{ for some } c_1 > 0.$$

So from (4.12), we infer that $\left\{ \frac{u_n}{\|x_n\|^{p-1}} \right\}_{n \geq 1} \subseteq L^q(Z)$ is bounded and by passing to a subsequence if necessary, we may assume that
$$\frac{u_n}{\|x_n\|^{p-1}} \xrightarrow{w} f_0 \text{ in } L^q(Z).$$
For every $\varepsilon > 0$ and $n \geq 1$, we introduce the sets
$$C_{\varepsilon,n}^+ = \left\{ z \in Z : x_n(z) > 0, \eta(z) - \varepsilon \leq \frac{u_n(z)}{(x_n(z))^{p-1}} \leq \widehat{\eta}(z) + \varepsilon \right\}$$

and
$$C_{\varepsilon,n}^{-} = \left\{z \in Z : x_n(z) < 0, \eta(z) - \varepsilon \leq \frac{u_n(z)}{|x_n(z)|^{p-2} x_n(z)} \leq \widehat{\eta}(z) + \varepsilon\right\}.$$

Since $\|x_n\| \to 0$, we may assume (at least for a subsequence) that
$$x_n(z) \to 0 \text{ a.e. on } Z \text{ as } n \to \infty.$$

Then by virtue of hypothesis $(H_j)^1(v)$, we have
$$\chi_{C_{\varepsilon,n}^+}(z) \to 1 \text{ a.e. on } \{v > 0\} \text{ and } \chi_{C_{\varepsilon,n}^-}(z) \to 1 \text{ a.e. on } \{v < 0\}.$$

Note that
$$\left\|\left(1 - \chi_{C_{\varepsilon,n}^+}\right) \frac{u_n(z)}{\|x_n\|^{p-1}}\right\|_{L^q(\{v>0\})} \to 0$$

and
$$\left\|\left(1 - \chi_{C_{\varepsilon,n}^-}\right) \frac{u_n(z)}{\|x_n\|^{p-1}}\right\|_{L^q(\{v<0\})} \to 0.$$

From the definition of the set $C_{\varepsilon,n}^+$, we have
$$\chi_{C_{\varepsilon,n}^+}(z) \frac{u_n(z)}{\|x_n\|^{p-1}} = \chi_{C_{\varepsilon,n}^+}(z) \frac{u_n(z)}{(x_n(z))^{p-1}} (v_n(z))^{p-1},$$

hence
$$\chi_{C_{\varepsilon,n}^+}(z)(\eta(z) - \varepsilon) \frac{u_n(z)}{(x_n(z))^{p-1}} (v_n(z))^{p-1}$$
$$\leq \chi_{C_{\varepsilon,n}^+}(z) \frac{u_n(z)}{\|x_n\|^{p-1}}$$
$$\leq \chi_{C_{\varepsilon,n}^+}(z)(\widehat{\eta}(z) + \varepsilon) \frac{u_n(z)}{(x_n(z))^{p-1}} (v_n(z))^{p-1} \text{ a.e. on } Z.$$

Taking weak limits in $L^q(\{v > 0\})$ and using Mazur's lemma, we obtain
$$(\eta(z) - \varepsilon)(v(z))^{p-1} \leq f_0(z) \leq (\widehat{\eta}(z) + \varepsilon)(v(z))^{p-1} \text{ a.e. on } \{v > 0\}.$$

Since $\varepsilon > 0$ was arbitrary, we let $\varepsilon \downarrow 0$ and get

(4.13) $\quad \eta(z)(v(z))^{p-1} \leq f_0(z) \leq \widehat{\eta}(z)(v(z))^{p-1} \text{ a.e. on } \{v > 0\}.$

In a similar fashion, working this time with the sets $\{C_{\varepsilon,n}^-\}_{\varepsilon > 0, n \geq 1}$, we obtain

(4.14) $\quad \widehat{\eta}(z)|v(z)|^{p-2} v(z) \leq f_0(z) \leq \eta(z)|v(z)|^{p-2} v(z) \text{ a.e. on } \{v < 0\}.$

Moreover, from (4.12) it is clear that

(4.15) $\quad f_0(z) = 0 \text{ a.e. on } \{v = 0\}.$

From (4.13), (4.14) and (4.15), it follows that

(4.16) $\quad f_0(z) = g_0(z)|v(z)|^{p-2} v(z) \text{ a.e. on } Z,$

with $g_0 \in L^\infty(Z)_+$ such that $\eta(z) \leq g_0(z) \leq \widehat{\eta}(z)$ a.e. on Z. Note that

(4.17) $\quad t_n \frac{u_n}{\|x_n\|^{p-1}} + (1 - t_n) hK(v_n) \to tg_0 K(v) + (1 - t) hK(v) \text{ in } W^{-1,q}(Z)$

and (see (4.9))
$$\text{(4.18)} \quad t_n \frac{u_n}{\|x_n\|^{p-1}} + (1-t_n)hK(v_n) \in A(v_n) + \partial \psi_{t_n\|x_n\|}(x_n) \text{ for all } n \geq 1.$$

We will show that $(t,x) \to A(x) + \partial \psi_t(x)$ is a pseudomonotone homotopy. To this end, let $t_n \to t$, $x_n \xrightarrow{w} x$ in $W_0^{1,p}(Z)$, $x_n^* \xrightarrow{w} x^*$ in $W^{-1,q}(Z)$ and
$$x_n^* \in A(x_n) + \partial \psi_{t_n}(x_n),$$
and suppose that
$$\text{(4.19)} \quad \limsup_{n \to \infty} \langle x_n^*, x_n - x \rangle \leq 0.$$

Since A is bounded, we may assume that
$$A(x_n) \xrightarrow{w} u^* \text{ in } W^{-1,q}(Z).$$

We know that $(t,x) \to \partial \psi_t(x)$ is a pseudomonotone homotopy (recall that $\psi_{t_n} \xrightarrow{M} \psi_t$ if $t_n \to t$ in $[0,1]$) and that $x_n^* - A(x_n) \in \partial \psi_{t_n}(x_n)$ for all $n \geq 1$. So, from Definition $1(b), (b_1)$, it follows that
$$\text{(4.20)} \quad x^* - u^* \in \partial \psi_t(x) \text{ and } \langle x_n^* - A(x_n), x_n \rangle \to \langle x^* - u^*, x \rangle.$$

Therefore, we have that
$$\lim_{n \to \infty} \langle x_n^* - A(x_n), x_n - x \rangle = 0.$$

Combining this with (4.19), we deduce that
$$\text{(4.21)} \quad \liminf_{n \to \infty} \langle A(x_n), x_n - x \rangle \leq 0.$$

But A is of type $(S)_+$. So, from (4.21) it follows that
$$\lim_{n \to \infty} \langle A(x_n), x_n - x \rangle = 0,$$
hence
$$x_n \to x \text{ in } W_0^{1,p}(Z),$$
and so
$$A(x_n) \xrightarrow{w} A(x) \text{ in } W^{-1,q}(Z),$$
i.e.,
$$u^* = A(x).$$

So, from (4.20), we have
$$x^* \in A(x) + \partial \psi_t(x) \text{ and } \langle x_n^*, x_n \rangle \to \langle x^*, x \rangle,$$
hence $(t,x) \to A(x) + \partial \psi_t(x)$ is a pseudomonotone homotopy (see Definition $1(b), (b_1)$). Then, from (4.17) and (4.18), and since $t_n\|x_n\| \to 0$ and $x_n \xrightarrow{w} x$ in $W_0^{1,p}(Z)$ we obtain
$$tg_0 K(v) + (1-t)hK(v) \in A(v) + \partial \psi_0(v) = A(v)$$
(since $\psi_0 = i_{W_0^{1,p}(Z)} \equiv 0$), hence
$$A(v) = \widehat{g}_t K(v) \text{ with } \widehat{g}_t = tg_0 + (1-t)h,$$
therefore
$$\text{(4.22)} \quad \begin{cases} -\operatorname{div}\left(\|Dv(z)\|_{\mathbb{R}^N}^{p-2} Dv(z)\right) = \widehat{g}_t(z)|v(z)|^{p-2}v(z) \text{ a.e. on } Z, \\ v(z) = 0 \text{ on } \partial Z. \end{cases}$$

Note that $\lambda_1 \leq \widehat{g}_t(z)$ a.e. on Z with strict inequality on a set of positive measure and
$$\widehat{g}_t(z) < \lambda_2 \text{ a.e. on } Z.$$
So, exploiting the strict monotonicity of $\widehat{\lambda}_1(m)$ and $\widehat{\lambda}_2(m)$ with respect to the weight function m, from (4.22) we infer that $v = 0$ (see also Kobayashi-Otani [**27**], Lemma 4.2). On the other hand, since A is of type $(S)_+$, we have

(4.23) $$\liminf_{n \to \infty} \langle A(v_n), v_n - v \rangle \geq 0.$$

Indeed, if this is not the case, we can find a subsequence $\{v_{n_k}\}_{k \geq 1}$ of $\{v_n\}_{n \geq 1}$ such that

(4.24) $$\lim_{k \to \infty} \langle A(v_{n_k}), v_{n_k} - v \rangle < 0,$$

hence
$$v_{n_k} \to v \text{ in } W_0^{1,p}(Z)$$
(since A is of type $(S)_+$, see Lemma 11), therefore
$$\lim_{k \to \infty} \langle A(v_{n_k}), v_{n_k} - v \rangle = 0,$$
a contradiction (see (4.24)). This proves (4.23). We have (cf. (4.9))
$$t_n \frac{u_n}{\|x_n\|^{p-1}} + (1 - t_n) hK(v_n) - A(v_n) \in \partial \psi_{t_n}(v_n), \ n \geq 1.$$

Then, because of (4.23) and since
$$\left\langle t_n \frac{u_n}{\|x_n\|^{p-1}} + (1 - t_n) hK(v_n), v_n - v \right\rangle$$
$$= \int_Z t_n \frac{u_n(z)}{\|x_n\|^{p-1}} + (1 - t_n) h(z) |v_n(z)|^{p-2} v_n(z) (v_n(z) - v(z)) \, dz \to 0,$$
we have
$$\limsup_{n \to \infty} \left\langle t_n \frac{u_n}{\|x_n\|^{p-1}} + (1 - t_n) hK(v_n) - A(v_n), v_n - v \right\rangle \leq 0.$$

As before, since A is bounded, we may assume that $A(v_n) \xrightarrow{w} u^*$ in $W^{-1,q}(Z)$. Then since $(t, x) \to \partial \psi_t(x)$ is a pseudomonotone homotopy and $\psi_0 \equiv 0$, we have (see Definition 1(b), (b_1))
$$\left\langle t_n \frac{u_n}{\|x_n\|^{p-1}} + (1 - t_n) hK(v_n) - A(v_n), v_n - v \right\rangle \to 0.$$

Then
$$\lim \langle A(v_n), v_n - v \rangle = 0$$
(recall that $\partial \psi_0 \equiv \{0\}$), hence
$$v_n \to v \text{ in } W_0^{1,p}(Z)$$
(since A is of type $(S)_+$, see Lemma 11). Therefore $\|v\| = 1$, in particular $v \neq 0$, a contradiction. This proves the Claim. The Claim permits the use of the homotopy invariance property of the degree map and so we have
$$\deg(A - N + \partial \psi, B_\rho(0), 0) = \deg_{(S)_+}(A - hK, B_\rho(0), 0)$$

for all $0 < \rho \leq \rho_0$ (recall that $\partial \psi_0 = \{0\}$). But from Drabek-Kufner-Nicolosi [15] (Chapter 3.6), we know that
$$\deg_{(S)_+} (A - hK, B_\rho(0), 0) = -1.$$
So, finally we have
$$\deg(A - N + \partial \psi, B_\rho(0), 0) = -1$$
for all $0 < \rho \leq \rho_0$. This completes the proof of Proposition 14. $\square$

Next, we will prove an analogous result for large balls. First, we identify another class of pseudomonotone homotopies.

LEMMA 15. *If X is a reflexive Banach space such that both X and X^* are locally uniformly convex, $A : X \to X^*$ is a bounded demicontinuous operator of type $(S)_+$ and $\psi \in \Gamma_0(X)$, then*
$$(t, x) \to h(t, x) = A(x) + t\partial \psi(x), \quad (t, x) \in T \times X$$
is a pseudomonotone homotopy.

PROOF. Let $t_n \to t$ in $[0,1]$, $x_n \xrightarrow{w} x$ in X, $x_n^* \xrightarrow{w} x^*$ in X^*, $x_n^* \in A(x_n) + t_n \partial \psi(x_n)$, $n \geq 1$, and
$$\limsup_{n \to \infty} \langle x_n^*, x_n - x \rangle \leq 0.$$
Note that $t_n \partial \psi(x_n) = \partial(t_n \psi)(x_n)$ and clearly $t_n \psi \xrightarrow{M} t\psi$. So
$$(t, x) \to \partial(t\psi)(x) = t\partial \psi(x),$$
is a pseudomonotone homotopy (see Definition 1$(b), (b_4)$). Then arguing as in the proof of Proposition 14, we conclude that $(t, x) \to A(t) + t\partial \psi(x)$ is a pseudomonotone homotopy. $\square$

PROPOSITION 16. *If hypotheses (H_γ) and $(H_j)^1$ hold, then there exists $R_0 > 0$ such that for all $R \geq R_0$, we have*
$$\deg(A - N + \partial \psi, B_R(0), 0) = 1.$$

PROOF. In this case, for $t \in [0,1]$, we define
$$C(t) = \left\{ y \in W_0^{1,p}(Z) : y(z) \geq t\gamma(z) \text{ a.e. on } Z \right\}.$$
Evidently, $C(1) = C$ and $C(0) = W_0^{1,p}(Z)_+$ (the positive cone of the ordered Banach space $W_0^{1,p}(Z)$). Also, let
$$\psi_t = i_{C(t)} \text{ for all } t \in [0,1].$$
If $t_n \to t$ in $[0,1]$, then, as before, we can easily check that
$$C(t) \xrightarrow{M} C(t)$$
and so
$$\psi_{t_n} \xrightarrow{M} \psi_t.$$
Therefore
$$(t, x) \to \partial \psi_t(x)$$
is a pseudomonotone homotopy. Also
$$\widehat{h}(t, x) = A(x) - tN(x) \text{ for } (t, x) \in [0, 1] \times W_0^{1,p}(Z),$$

(where $N(x) = \{u \in L^q(Z) : u(z) \in \partial j(z, x(z))$ a.e. on $Z\}$) is an $(S)_+$ homotopy. Therefore
$$h(t, x) = A(x) - tN(x) + \partial \psi_t(x), \ (t, x) \in [0, 1] \times W_0^{1,p}(Z),$$
is an admissible homotopy.

Claim: There exists $R_0 \geq 0$ such that for all $t \in [0, 1]$, all $R \geq R_0$ and all $x \in \partial B_R(0)$, we have
$$0 \notin h(t, x) = A(x) - tN(x) + \partial \psi_t(x).$$

Again, we argue indirectly. So, suppose that the Claim is not true. Then we can find $\{t_n\}_{n \in \mathbb{N}} \subseteq [0, 1]$ and $x_n \in C(t_n)$, $n \geq 1$, such that
$$t_n \to t \text{ in } [0, 1], \|x_n\| \to \infty \text{ and } 0 \in h(t_n, x_n), \ n \geq 1.$$

We have
$$-A(x_n) + t_n u_n \in \partial \psi_{t_n}(x_n) \text{ for some } u_n \in N(x_n), \ \forall n \geq 1,$$
hence
$$(4.25) \quad \langle A(x_n), y - x_n \rangle - t_n \int_Z u_n(z)(y(z) - x_n(z)) \, dz \geq 0 \text{ for all } y \in C(t_n).$$

Set $v_n = \frac{x_n}{\|x_n\|}$, $n \geq 1$. We may assume that
$$v_n \xrightarrow{w} v \text{ in } W_0^{1,p}(Z) \text{ and } v_n \to v \text{ in } L^p(Z).$$

We can easily see that $C(t_n) = \|x_n\| C\left(\frac{t_n}{\|x_n\|}\right)$, $n \geq 1$. Then $v_n \in C\left(\frac{t_n}{\|x_n\|}\right)$ and so
$$v \in C(0) = W_0^{1,p}(Z)_+.$$

Also, if we divide (4.25) by $\|x_n\|^p$, we obtain
$$(4.26) \quad \langle A(v_n), \widehat{y} - v_n \rangle - t_n \int_Z \frac{u_n(z)}{\|x_n\|^{p-1}}(\widehat{y}(z) - v_n(z)) \, dz \geq 0$$

for all $\widehat{y} \in C\left(\frac{t_n}{\|x_n\|}\right)$. From (4.12) we see that $\left\{\frac{u_n}{\|x_n\|^{p-1}}\right\}_{n \geq 1} \subseteq L^q(Z)$ is bounded. So, we may assume that
$$\frac{u_n}{\|x_n\|^{p-1}} \xrightarrow{w} f_\infty \text{ in } L^q(Z), \text{ as } n \to \infty.$$

For every $\varepsilon > 0$ and $n \geq 1$, we introduce the set
$$D_{\varepsilon,n}^+ = \left\{z \in Z : x_n(z) > 0, -\varepsilon \leq \frac{u_n(z)}{x_n(z)^{p-1}} \leq \theta(z) + \varepsilon\right\}.$$

Because of hypothesis $(H_j)^1(iv)$, we see that
$$\chi_{D_{\varepsilon,n}^+}(z) \to 1 \text{ a.e. on } \{v > 0\}.$$

Note that
$$\left\|\left(1 - \chi_{D_{\varepsilon,n}^+}(z)\right) \frac{u_n}{\|x_n\|^{p-1}}\right\|_{L^q(\{v>0\})} \to 0,$$
hence
$$\chi_{D_{\varepsilon,n}^+}(z) \frac{u_n}{\|x_n\|^{p-1}} \xrightarrow{w} f_\infty \text{ in } L^q(\{v > 0\}).$$

From the definition of $D_{\varepsilon,n}^+$, we see that

$$\chi_{D_{\varepsilon,n}^+}(z)(-\varepsilon)(v_n(z))^{p-1} \leq \chi_{D_{\varepsilon,n}^+}(z)\frac{u_n(z)}{\|x_n\|^{p-1}}$$

$$= \chi_{D_{\varepsilon,n}^+}(z)\frac{u_n(z)}{(x_n(z))^{p-1}}(v_n(z))^{p-1}$$

$$\leq \chi_{D_{\varepsilon,n}^+}(z)(\theta(z)+\varepsilon)(v_n(z))^{p-1} \text{ a.e. on } Z.$$

Taking weak limits in $L^q(\{v > 0\})$ and using Mazur's lemma, we obtain

$$-\varepsilon(v(z))^{p-1} \leq f_\infty(z) \leq (\theta(z)+\varepsilon)(v(z))^{p-1} \text{ a.e. on } \{v > 0\}.$$

Let $\varepsilon \downarrow 0$, to arrive at

$$0 \leq f_\infty(z) \leq \theta(z)(v(z))^{p-1} \text{ a.e. on } \{v > 0\}.$$

On the other hand, from (4.12), we see that

$$f_\infty(z) = 0 \text{ a.e. on } \{v = 0\}.$$

Since $Z = \{v > 0\} \cup \{v = 0\}$ (recall that $v \in W_0^{1,p}(Z)_+$), we have

$$0 \leq f_\infty(z) \leq \theta(z)(v(z))^{p-1} \text{ a.e. on } Z,$$

therefore

$$f_\infty = g_\infty v^{p-1} \text{ with } g_\infty \in L^\infty(Z)_+, g_\infty(z) \leq \theta(z) \text{ a.e. on } Z.$$

Since $v \in W_0^{1,p}(Z)_+$, we have $v \in C\left(\frac{t_n}{\|x_n\|}\right)$ and so, in (4.26), we can set $\widehat{y} = v$ to obtain

$$\langle A(v_n), v_n - v \rangle \leq t_n \int_Z \frac{u_n(z)}{\|x_n\|^{p-1}}(v_n(z) - v(z))dz,$$

hence

$$\limsup_{n\to\infty} \langle A(v_n), v_n - v \rangle \leq 0,$$

and since A is of type $(S)_+$ (see Lemma 11),

$$v_n \to v \text{ in } W_0^{1,p}(Z).$$

Recall that

$$C\left(\frac{t_n}{\|x_n\|}\right) \xrightarrow{M} C(0) = W_0^{1,p}(Z)_+.$$

So, for any $\widehat{y}_0 \in W_0^{1,p}(Z)_+$, we can find $\widehat{y}_n \in C\left(\frac{t_n}{\|x_n\|}\right)$, $n \geq 1$, such that

$$\widehat{y}_n \to \widehat{y}_0 \text{ in } W_0^{1,p}(Z).$$

We set $\widehat{y} = \widehat{y}_n$ in (4.26) and then pass to the limit as $n \to \infty$. We obtain
(4.27)
$$\langle A(v), \widehat{y}_0 - v \rangle \geq t \int_Z g_\infty(z)(v(z))^{p-1}(\widehat{y}_0(z) - v(z))dz, \forall \widehat{y}_0 \in W_0^{1,p}(Z)_+.$$

Let $w \in W_0^{1,p}(Z)$, $\varepsilon > 0$ and consider
(4.28)
$$(v + \varepsilon w)^+ = v + \varepsilon w + (v + \varepsilon w)^-,$$

where $u^+ := \max\{0, u\}$ and $u^- := -\min\{0, u\}$. If $x^* = A(v) - tg_\infty K(v)$ and we take

$$\widehat{y}_0 = (v + \varepsilon w)^+ \in C(0) = W_0^{1,p}(Z)_+,$$

in (4.27), we get
$$\langle x^*, (v+\varepsilon w)^+ - v\rangle \geq 0,$$
hence

(4.29) $$\langle x^*, \varepsilon w\rangle \geq -\langle x^*, (v+\varepsilon w)^-\rangle.$$

Note that
$$-\langle x^*, (v+\varepsilon w)^-\rangle = -\langle A(v), (v+\varepsilon w)^-\rangle$$
(4.30) $$+ t\int_Z g_\infty(z)(v(z))^{p-1}(v(z)+\varepsilon w(z))^- dz.$$

We set
$$Z_-^\varepsilon := \{v+\varepsilon w < 0\}.$$
Then we have
$$-\langle A(v), (v+\varepsilon w)^-\rangle = -\int_Z \|Dv\|_{\mathbb{R}^N}^{p-2}\left(Dv, D(v+\varepsilon w)^-\right)_{\mathbb{R}^N} dz$$
$$= \int_{Z_-^\varepsilon} \|Dv\|_{\mathbb{R}^N}^{p-2}(Dv, D(v+\varepsilon w))_{\mathbb{R}^N} dz$$
(4.31) $$\geq \varepsilon \int_{Z_-^\varepsilon} \|Dv\|_{\mathbb{R}^N}^{p-2}(Dv, Dw)_{\mathbb{R}^N} dz.$$

Also we have

(4.32) $$\int_Z g_\infty v^{p-1}(v+\varepsilon w)^- dz \geq 0.$$

(since $g_\infty, v \geq 0$). Using (4.31) and (4.32) in (4.30), we obtain

(4.33) $$-\langle x^*, (v+\varepsilon w)^-\rangle \geq \varepsilon \int_{Z_-^\varepsilon} \|Dv\|_{\mathbb{R}^N}^{p-2}(Dv, Dw)_{\mathbb{R}^N} dz.$$

From (4.29) and (4.33) it follows that
$$\varepsilon \langle x^*, w\rangle \geq \varepsilon \int_{Z_-^\varepsilon} \|Dv\|_{\mathbb{R}^N}^{p-2}(Dv, Dw)_{\mathbb{R}^N} dz,$$
hence

(4.34) $$\langle x^*, w\rangle \geq \int_{Z_-^\varepsilon} \|Dv\|_{\mathbb{R}^N}^{p-2}(Dv, Dw)_{\mathbb{R}^N} dz.$$

Note that $Z_-^\varepsilon \to \{v=0\}$ as $\varepsilon \downarrow 0$ and (recall that $v \in W_0^{1,p}(Z)$)
$$Dv(z) = 0 \text{ a.e. on } \{v=0\}.$$
So, if we pass to the limit as $\varepsilon \downarrow 0$ in (4.34), we obtain
$$\langle x^*, w\rangle \geq 0.$$
But $w \in W_0^{1,p}(Z)$ was arbitrary. So it follows that $x^* = 0$, hence
$$A(v) = tg_\infty v^{p-1},$$
therefore

(4.35) $$\begin{cases} -div\left(\|Dv(z)\|_{\mathbb{R}^N}^{p-2} Dv(z)\right) = tg_\infty(z)|v(z)|^{p-2}v(z) \text{ a.e. on } Z, \\ v(z) = 0 \text{ on } \partial Z. \end{cases}$$

Since $\|v\| = 1$, we have $v \neq 0$ and so v is an eigenfunction of the weighted eigenvalue problem (4.35), with weight $tg_\infty \in L^\infty(Z)_+$. Exploiting the strict monotonicity on the weight of the principal eigenvalue and since
$$0 \leq tg_\infty \leq g_\infty \leq \theta,$$
we have
$$\widehat{\lambda}_1(tg_\infty) \geq \widehat{\lambda}_1(g_\infty) \geq \widehat{\lambda}_1(\theta) > \widehat{\lambda}_1(\lambda_1) = 1.$$
Then from (4.35) we infer that $v = 0$, a contradiction. This proves the Claim. As before, the homotopy invariance of the degree map implies that
(4.36) $$\deg(A - N + \partial\psi, B_R(0), 0)$$
$$= \deg(A + \partial\psi_0, B_R(0), 0) \text{ for all } R \geq R_0.$$
Let
$$h_1(t, x) = A(x) + t\partial\psi_0(x), \ (t, x) \in [0, 1] \times W_0^{1,p}(Z).$$
From Lemma 15, we know that $h_1(.,.)$ is a pseudomonotone homotopy. Next we show that if $R_0 > 0$ is chosen large enough, then for all $t \in [0, 1]$, all $R \geq R_0$ and all $x \in \partial B_R(0)$, we have
$$0 \notin h_1(t, x).$$
As before, suppose that we can find $\{t_n\}_{n \geq 1} \subseteq [0, 1]$ and $\{x_n\}_{n \geq 1} \subseteq W_0^{1,p}(Z)$ such that $t_n \to t$ in $[0, 1]$, $\|x_n\| \to \infty$ and
$$0 \in h_1(t_n, x_n) \text{ for all } n \geq 1.$$
Then we have
$$0 \in A(x_n) + t_n \partial\psi_0(x_n) \text{ for all } n \geq 1,$$
hence
$$\langle A(x_n), y - x_n \rangle \geq 0 \text{ for all } y \in C(0) = W_0^{1,p}(Z)_+.$$
Dividing by $\|x_n\|^p$ and setting $v_n = \frac{x_n}{\|x_n\|}$, $n \geq 1$, we obtain
(4.37) $$\langle A(v_n), y - v_n \rangle \geq 0 \text{ for all } y \in C(0) = W_0^{1,p}(Z)_+.$$
We may assume that
$$v_n \xrightarrow{w} v \text{ in } W_0^{1,p}(Z) \text{ and } v_n \to v \text{ in } L^p(Z), \text{ with } v \in W_0^{1,p}(Z)_+.$$
We set $y = v$ in (4.37). Hence
$$\langle A(v_n), v_n - v \rangle \leq 0 \text{ for all } n \geq 1.$$
Since A is of type $(S)_+$, it follows that
$$v_n \to v \text{ in } W_0^{1,p}(Z),$$
hence, $\|v\| = 1$, and therefore
$$v \neq 0.$$
Passing to the limit as $n \to \infty$ in (4.37), we obtain
$$\langle A(v), y - v \rangle \geq 0 \text{ for all } y \in W_0^{1,p}(Z)_+.$$
Let $w \in W_0^{1,p}(Z)$, $\varepsilon > 0$ and
$$y = (v + \varepsilon w)^+ = v + \varepsilon w + (v + \varepsilon w)^-.$$
Then, as before, we have
$$\langle A(v), w \rangle \geq 0$$

hence $A(v) = 0$, and consequently,
$$v = 0,$$
a contradiction. Then, from the homotopy invariance of the degree, we have
(4.38) $\qquad \deg(A + \partial \psi_0, B_R(0), 0) = \deg_{(S)_+}(A, B_R(0), 0)$ for all $R \geq R_0$.

Now consider the $(S)_+$–homotopy
$$h_2(t, x) = tA(x) + (1-t)\mathcal{F}(x) \text{ for all } (t, x) \in [0, 1] \times W_0^{1,p}(Z).$$
Clearly $\langle h_2(t, x), x \rangle \neq 0$ for all $x \neq 0$ and so, by the homotopy invariance of $\deg_{(S)_+}$, we have
(4.39) $\qquad \deg_{(S)_+}(A, B_R(0), 0) = \deg_{(S)_+}(\mathcal{F}, B_R(0), 0) = 1$
(see Chapter 2). From (4.36), (4.38) and (4.39), we finally conclude that
$$\deg(A - N + \partial \psi, B_R(0), 0) = 1$$
for all $R \geq R_0$. $\qquad \square$

Now we can state the multiplicity result for problem (4.1).

THEOREM 17. *If hypotheses (H_γ) and $(H_j)^1$ hold, then the problem (4.1) has at least two nontrivial solutions $x_0, \widehat{x} \in W_0^{1,p}(Z)$.*

PROOF. From Proposition 13, we know that we can find $x_0 \in W_0^{1,p}(Z)$ such that
(4.40) $\qquad\qquad\qquad \varphi(x_0) = \inf_{x \in W_0^{1,p}(Z)} \varphi(x).$

Invoking Theorem 8, we can find $r > 0$ such that
(4.41) $\qquad\qquad\qquad \deg(A - N + \partial \psi, B_r(x_0), 0) = 1.$

From (4.41) and Proposition 14, it follows that $x_0 \neq 0$. We choose $\rho_0 > 0$ small so that
$$B_r(x_0) \cap B_{\rho_0}(0) = \varnothing$$
and $R_0 > 0$ large such that
$$B_{\rho_0}(0), B_r(x_0) \subseteq B_{R_0}(0).$$
From the aditivity of the domain property of the degree map, we have
$$\deg(A - N + \partial \psi, B_{R_0}(0), 0) = \deg(A - N + \partial \psi, B_r(0), 0)$$
$$+ \deg(A - N + \partial \psi, B_{\rho_0}(0), 0)$$
$$+ \deg(A - N + \partial \psi, B_{R_0}(0) \setminus (B_r(x_0) \cup B_{\rho_0}(0)), 0)$$
hence
$$1 = \deg(A - N + \partial \psi, B_{R_0}(0) \setminus (B_r(x_0) \cup B_{\rho_0}(0)), 0)$$
(see Theorem 8 and Propositions 14 and 16). From the existence property of the degree map, it follows that there exists
$$\widehat{x} \in B_{R_0}(0) \setminus (B_r(x_0) \cup B_{\rho_0}(0))$$
hence $\widehat{x} \neq x_0$, $\widehat{x} \neq 0$, such that
$$0 \in A(\widehat{x}) - N(\widehat{x}) + \partial \psi(\widehat{x}) = \partial \widehat{\varphi}(\widehat{x}) + \partial \psi(\widehat{x}),$$

i.e.,
$$-A(\widehat{x}) + u \in \partial \psi(\widehat{x}) \text{ for some } u \in N(\widehat{x}).$$
The last relation implies
$$\langle A(\widehat{x}), y - \widehat{x} \rangle - \int_Z u(z)(y(z) - \widehat{x}(z)) dz \geq 0 \text{ for all } y \in C,$$
therefore $\widehat{x} \in W_0^{1,p}(Z)$ is a nontrivial solution of (4.1). On the other hand, by (4.40), for all $\lambda > 0$ and all $y \in W_0^{1,p}(Z)$ one has
$$0 \leq \varphi(x_0 + \lambda y) - \varphi(x_0) = \widehat{\varphi}(x_0 + \lambda y) - \widehat{\varphi}(x_0) + \psi(x_0 + \lambda y) - \psi(x_0)$$
hence
$$0 \leq \frac{1}{\lambda}(\widehat{\varphi}(x_0 + \lambda y) - \widehat{\varphi}(x_0)) + \frac{1}{\lambda}(\psi(x_0 + \lambda y) - \psi(x_0))$$
$$\leq \frac{1}{\lambda}[\widehat{\varphi}(x_0 + \lambda y) - \widehat{\varphi}(x_0)] + (\psi(x_0 + y) - \psi(x_0))$$
(since ψ is convex). Passing to the limit as $\lambda \downarrow 0$, we obtain
(4.42)
$$0 \leq \widehat{\varphi}^0(x_0; y) + \psi(x_0 + y) - \psi(x_0).$$
We set
$$\phi_1(y) := \widehat{\varphi}^0(x_0; y)$$
and
$$\phi_2(y) = \psi(x_0 + y) - \psi(x_0).$$
Evidently, ϕ_1 is continuous and convex (in fact sublinear) while $\phi_2 \in \Gamma_0\left(W_0^{1,p}(Z)\right)$. Also
$$\partial \phi_1(0) = \partial \widehat{\varphi}(x_0)$$
(where the first is a convex subdifferential and the second is a generalized one) and
$$\partial \phi_2(0) = \partial \psi(x_0)$$
(both are convex subdifferentials). From convex analysis we know that
(4.43) $\quad \partial(\phi_1 + \phi_2)(0) = \partial \phi_1(0) + \partial \phi_2(0) = \partial \widehat{\varphi}(x_0) + \partial \psi(x_0)$
(see Denkowski-Migorski-Papageorgiou [13], p.549). Since
$$\phi_1(0) = \phi_2(0) = 0$$
from (4.42) we see that
$$0 \in \partial(\phi_1 + \phi_2)(0)$$
hence, by (4.43),
$$0 \in \partial \widehat{\varphi}(x_0) + \partial \psi(x_0).$$
This implies that
$$-A(x_0) + u \in \partial \psi(x_0) \text{ for some } u \in N(x_0),$$
hence
$$\langle A(x_0), y - x_0 \rangle - \int_Z u(z)(y(z) - x_0(z)) dz \geq 0 \text{ for all } y \in C.$$
Consequently $x_0 \in W_0^{1,p}(Z)$ is a second nontrivial solution of (4.1). $\square$

If the obstacle function γ is identically zero, then we can weaken the hypothesis $(H_j)^1(v)$. Now, the constraint set of the unilateral problem is
$$C = \left\{ y \in W_0^{1,p}(Z) : y(z) \geq 0 \right\}.$$
The hypotheses on the nonsmooth potential $j(z, x)$ are now the following:

$(H_j)^2$ $j : Z \times \mathbb{R} \to \mathbb{R}$ is a function such that that $j(z, 0) = 0$ a.e. on Z and
(i) for all $x \in \mathbb{R}$, $z \to j(z, x)$ is measurable;
(ii) for almost all $z \in Z$, $x \to j(z, x)$ is locally Lipschitz;
(iii) for almost all $z \in Z$, all $x \in \mathbb{R}$ and all $u \in \partial j(z, x)$, we have
$$|u| \leq a(z) + c|x|^{p-1}, \text{ with } a \in L^\infty(Z)_+, c > 0;$$
(iv) there exists $\theta \in L^\infty(Z)_+$ such that $\theta(z) \leq \lambda_1$ a.e. on Z with strict inequality on a set of positive measure ($\lambda_1 > 0$ is the principal eigenvalue of $\left(-\triangle_p, W_0^{1,p}(Z)\right)$ with weight $m = 1$) and
$$0 \leq \liminf_{x \to +\infty} \frac{u}{x^{p-1}} \leq \limsup_{x \to +\infty} \frac{u}{x^{p-1}} \leq \theta(z)$$
uniformly for almost all $z \in Z$ and all $u \in \partial j(z, x)$.
(v) there exist $\eta, \widehat{\eta} \in L^\infty(Z)_+$ such that $\lambda_1 \leq \eta(z)$ a.e. on Z with strict inequality on a set of positive measure and
$$\eta(z) \leq \liminf_{x \to 0^+} \frac{u}{x^{p-1}} \leq \limsup_{x \to 0^+} \frac{u}{x^{p-1}} \leq \widehat{\eta}(z)$$
uniformly for almost all $z \in Z$ and all $u \in \partial j(z, x)$.

REMARK 18. *In this case, near 0^+, we have only a nonuniform nonresonance condition from above with respect to $\lambda_1 > 0$. No condition with respect to λ_2 is imposed (compare with $(H_j)^1(v)$). So, the "slope" $\frac{u}{x^{p-1}}$ may cross any finite set of eigenvalues as we move from 0^+ to $+\infty$.*

In obtaining a multiplicity result for the new problem, only Proposition 14 changes. In fact, the statement of the result remains the same and only the proof changes.

PROPOSITION 19. *If hypothesis $(H_j)^2$ hold and $\gamma \equiv 0$, then there exists $\rho_0 > 0$ such that for all $0 < \rho \leq \rho_0$ we have*
$$\deg\left(\partial\widehat{\varphi} + \partial\psi, B_\rho(0), 0\right) = -1.$$

PROOF. Keeping the notation introduced in the proof of Proposition 14, we consider the homotopy
$$\widetilde{h}(t, x) = A(x) - tN(x) - (1 - t)hK(x) + \partial\psi(x)$$
for $(t, x) \in [0, 1] \times W_0^{1,p}(Z)$. Since
$$(t, x) \to A(x) - tN(x) - (1 - t)hK(x)$$
is an $(S)_+$− homotopy, we see that $(t, x) \to \widetilde{h}(t, x)$ is an admissible homotopy.
Claim: There exists $\rho_0 > 0$ such that for all $t \in [0, 1]$, all $0 < \rho \leq \rho_0$ and all $x \in \partial B_\rho(0)$ we have
$$0 \notin \widetilde{h}(t, x).$$

Suppose that the Claim is not true. Then we can find $\{t_n\}_{n\geq 1} \subseteq [0,1]$ and $x_n \in C$, $n \geq 1$, such that
$$t_n \to t \text{ in } [0,1], \ \|x_n\| \to 0$$
and
$$0 \in A(x_n) - t_n N(x_n) - (1-t_n) hK(x_n) + \partial\psi(x_n), \ n \geq 1.$$
We set
$$v_n = \frac{x_n}{\|x_n\|}, \ n \geq 1,$$
and by passing to a suitable subsequence if necessary, we may assume that
$$v_n \xrightarrow{w} v \text{ in } W_0^{1,p}(Z) \text{ and } v_n \to v \text{ in } L^p(Z).$$
Then
$$-A(x_n) + t_n u_n + (1-t_n) hK(x_n) \in \partial\psi(x_n), \text{ with } u_n \in N(x_n),$$
hence for all $y \in C$,
$$\langle A(x_n), y - x_n \rangle - t_n \int_Z u_n (y - x_n) dz + (1-t_n) \int_Z h |x_n|^{p-2} x_n (y - x_n) dz \geq 0.$$
Dividing this inequality by $\|x_n\|^p$, we obtain

(4.44)
$$\begin{aligned}&\langle A(v_n), y - v_n \rangle - t_n \int_Z \frac{u_n}{\|x_n\|^{p-1}} (y - v_n) dz \\ &- (1-t_n) \int_Z h |v_n|^{p-2} v_n (y - v_n) dz \geq 0.\end{aligned}$$

Arguing as in the proof of Proposition 14, we obtain
$$\frac{u_n}{\|x_n\|^{p-1}} \xrightarrow{w} f_0 \text{ in } L^p(Z),$$
with
$$f_0(z) = g_0(z) |v(z)|^{p-2} v(z) \text{ a.e. on } Z,$$
where $g_0 \in L^\infty(Z)_+$ is such that
$$\eta(z) \leq g_0(z) \leq \widehat{\eta}(z) \text{ a.e. on } Z.$$
Moreover, if (4.44) we set $y = v$, then since
$$\int_Z \frac{u_n(z)}{\|x_n\|^{p-1}} (v_n(z) - v(z)) dz \to 0,$$
and
$$\int_Z h(z) |v_n(z)|^{p-2} v_n(z) (v(z) - v_n(z)) dz \to 0,$$
from (4.44) we obtain
$$\limsup_{n \to \infty} \langle A(v_n), v_n - v \rangle \leq 0,$$
hence
$$v_n \to v \text{ in } W_0^{1,p}(Z).$$
So, if we pass to the limit as $n \to \infty$ in (4.44), we obtain
$$\langle A(v), y - v \rangle - t \int_Z g_0 |v|^{p-2} v (y - v) dz - (1-t) \int_Z h |v|^{p-2} v (y - v) dz \geq 0$$
for all $y \in C$. Set
$$\widehat{g}_t = t g_0 + (1-t) h.$$

We have
$$\langle A(v), y - v \rangle - \int_Z \widehat{g}_t(z)(v(z))^{p-1}(y(z) - v(z))\,dz \geq 0 \text{ for all } y \in C.$$

Using as before the test function $y = (v + \varepsilon w)^+$ for any $w \in W_0^{1,p}(Z)$ and $\varepsilon > 0$, finally we have

(4.45) $\begin{cases} -\mathrm{div}\left(\|Dv(z)\|_{\mathbb{R}^N}^{p-2} Dv(z)\right) = \widehat{g}_t(z)|v(z)|^{p-2}v(z) \text{ a.e. on } Z \\ v(z) = 0 \text{ on } \partial Z, \|v\| = 1. \end{cases}$

We know that
$$\widehat{g}_t(z) \geq \eta(z) \text{ a.e. on } Z$$
and so
$$\widehat{\lambda}_1(\widehat{g}_t) \leq \widehat{\lambda}_1(\eta) < \widehat{\lambda}_1(\lambda_1) = 1.$$

So from (4.45) it follows that $v \neq 0$ cannot be the principal eigenfunction of the weighted eigenvalue problem with weight $\widehat{g}_t \in L^\infty(Z)_+$. So, v must change sign, a contradiction to the fact that $v \in C$. Hence, the claim is true. Now the rest of the proof proceeds as the corresponding part of the proof of Proposition 14. So, finally, we have
$$\deg(A - N + \partial \psi, B_\rho(0), 0) = -1 \text{ for all } 0 < \rho \leq \rho_0,$$
ending the proof. $\square$

From Propositions 13, 16 and 19 we now obtain the following multiplicity result for problem (4.1) where the constraint set C is $W_0^{1,p}(Z)_+$. The proof is similar to that of Theorem 17 and so is omitted.

THEOREM 20. *If hypotheses $(H_j)^2$ hold and $\gamma \equiv 0$, then the problem (4.1) has at least two nontrivial solutions $x_0, \widehat{x} \in W_0^{1,p}(Z)$.*

In a similar fashion, we can also treat the following unilateral problem

(4.46) $\begin{cases} \int_Z \|Dx(z)\|_{\mathbb{R}^N}^{p-2}(Dx(z), Dy(z) - Dx(z))_{\mathbb{R}^N}\,dz \\ \geq \int_Z u(z)(y(z) - x(z))\,dz \text{ for all } y \in C \text{ with} \\ C = \left\{y \in W_0^{1,p}(Z) : \|Dy(z)\|_{\mathbb{R}^N} \leq c \text{ a.e. on } Z\right\}, c > 0, \\ u \in L^q(Z), u(z) \in \partial j(z, x(z)) \text{ a.e. on } Z \text{ and} \\ 1 < p < \infty, (\frac{1}{p} + \frac{1}{q} = 1). \end{cases}$

This problem was also studied by Kobayashi-Otani [27], but under the assumption that the potential is smooth. In contrast, here the potential function is in general nonsmooth and the nonuniform nonresonance condition that we use is more general, as we idicate below.

The hypotheses on the potential function $j(z, x)$ are:

$(H_j)^3$ $j : Z \times \mathbb{R} \to \mathbb{R}$ is a function such that that $j(z, 0) = 0$ a.e. on Z and
(i) for all $x \in \mathbb{R}$, $z \to j(z, x)$ is measurable;
(ii) for almost all $z \in Z$, $x \to j(z, x)$ is locally Lipschitz;

(iii) for every $r > 0$, there exists $a_r \in L^q(Z)_+$ such that for almost all $z \in Z$, all $x \in \mathbb{R}$ with $|x| \leq r$ and all $u \in \partial j(z, x)$, we have
$$|u| \leq a_r(z);$$

(iv) there exist functions $\eta, \widehat{\eta} \in L^\infty(Z)_+$ such that $\lambda_1 \leq \eta(z)$ a.e. on Z, with strict inequality on a set of positive measure, $\widehat{\eta}(z) < \lambda_2$ a.e. on Z and
$$\eta(z) \leq \liminf_{x \to 0} \frac{u}{|x|^{p-2}x} \leq \limsup_{x \to 0} \frac{u}{|x|^{p-2}x} \leq \widehat{\eta}(z)$$

uniformly for almost all $z \in Z$ and all $u \in \partial j(z, x)$.

REMARK 21. *Note that in $(H_j)^3$ (iv) we do not require that the limit of the "slope" $\frac{u}{|x|^{p-2}x}$ exists as $x \to 0$. So, $(H_j)^3$ (iv) is more general than the corresponding condition in Kobayashi-Otani* [27] *(in addition, in* [27] *the potential is smooth). The following smooth potential function satisfies our hypotheses with $p = 3$, but not those of Kobayashi-Otani* [27]*. For simplicity we drop the z-dependence.*

Let $\mu > 0$ be such that $\mu < \frac{1}{2}(\lambda_2 - \lambda_1)$ and $\lambda_1 + \mu < \lambda_2 - \mu$. Also let $\xi > 0$ be such that
$$\lambda_1 + \mu < \xi < \lambda_2 - \mu.$$

Consider the function
$$j(x) = \frac{\mu}{3} \sin x^3 + \frac{\xi}{3} |x|^3.$$

Then we have $j \in C^1(\mathbb{R})$ and
$$\lambda_1 < \xi - \mu = \liminf_{x \to 0} \frac{j'(x)}{|x|\, x} < \limsup_{x \to 0} \frac{j'(x)}{|x|\, x} = \xi + \mu < \lambda_2.$$

The Euler functional for problem (4.46) is given by
$$\varphi(x) = \widehat{\varphi}(x) + \psi(x) \text{ for } x \in W_0^{1,p}(Z),$$

where
$$\widehat{\varphi}(x) = \frac{1}{p} \|Dx\|_p^p - \int_Z j(z, x(z))\, dz$$

and
$$\psi(x) = i_C(x)$$

for $x \in W_0^{1,p}(Z)$. In this case the situation for large balls is much simpler (note that $(H_j)^3$ does not include any asymptotic conditions at $\pm\infty$).

PROPOSITION 22. *If hypotheses $(H_j)^3$ hold, then there exists $R_0 > 0$ such that for all $R \geq R_0$ we have*
$$\deg(\partial \widehat{\varphi} + \partial \psi, B_R(0), 0) = 1.$$

PROOF. Note that the constraint set C is bounded in $x \in W_0^{1,p}(Z)$ (by Poincaré's inequality). So, we can find $R_0 > 0$ such that
$$C \subseteq B_{R_0}(0).$$

Therefore, we infer that
$$0 \notin A(x) - tN(x) + \partial \psi(x)$$

for all $t \in [0,1]$ and all $x \in \partial B_{R_0}(0)$, where A and N are defined as before. Recall that the homotopy
$$(t,x) \to A(x) - tN(x) + \partial \psi(x)$$
is admissible. So, from the homotopy invariance of the degree map, we have

(4.47) $\qquad \deg(A - N + \partial \psi, B_R(0), 0) = \deg(A + \partial \psi, B_R(0), 0).$

But then, as in the proof of Proposition 16, we can check that
$$\deg(A + \partial \psi, B_R(0), 0) = \deg_{(S)_+}(\mathcal{F}, B_R(0), 0) = 1,$$
hence (cf. (4.47))
$$\deg(\partial \widehat{\varphi} + \partial \psi, B_R(0), 0) = 1 \text{ for all } R \geq R_0.$$
The proof is complete. $\qquad \square$

PROPOSITION 23. *If conditions $(H_j)^3$ hold, then there exists $\rho_0 > 0$ such that*
$$\deg(\partial \widehat{\varphi} + \partial \psi, B_\rho(0), 0) = -1 \text{ for all } 0 < \rho \leq \rho_0.$$

PROOF. For every $t \in (0,1]$, we set
$$C(t) = \left\{ y \in W_0^{1,p}(Z) : \|Dy(z)\|_{\mathbb{R}^N} \leq \frac{c}{t} \text{ a.e. on } Z \right\}$$
and $C(0) = W_0^{1,p}(Z)$. Also let
$$\psi_t = i_{C(t)} \in \Gamma_0\left(W_0^{1,p}(Z)\right).$$
For $h \in L^\infty(Z)_+$ such that
$$\eta(z) \leq h(z) \leq \widehat{\eta}(z) \text{ a.e. on } Z,$$
we consider the homotopy
$$h^*(t,x) = A(x) - tN(x) - (1-t)hK(x) + \partial \psi_t(x).$$
From the proof of Proposition 14, we know that $h^*(t,x)$ is an admissible homotopy. We will show that we can find $\rho_0 > 0$ such that
$$0 \notin A(x) - tN(x) - (1-t)hK(x) + \partial \psi_t(x)$$
for all
$$(t,x) \in [0,1] \times \partial B_\rho(0), \rho \leq \rho_0.$$
Suppose that this is not the case. Then we can find $t_n \in [0,1]$, $x_n \in C(t_n)$, $n \geq 1$, such that
$$t_n \to t \in [0,1], \|x_n\| \to 0$$
and

(4.48) $\qquad 0 \in A(x_n) - t_n N(x_n) - (1-t_n)hK(x_n) + \partial \psi_{t_n}(x_n).$

Recall that $C \subseteq B_{R_0}(0)$. So by truncating $j(z,.)$ at R_0, we may assume without any loss of generality that for almost all $z \in Z$, all $x \in \mathbb{R}$ and all $u \in \partial j(z,x)$, we have
$$|u| \leq a(z) \text{ with } a \in L^q(Z)_+.$$
We let
$$v_n = \frac{x_n}{\|x_n\|}, n \geq 1.$$
Then
$$v_n \xrightarrow{w} v \text{ in } W_0^{1,p}(Z) \text{ and } v_n \to v \text{ in } L^p(Z),$$

Dividing (4.48) by $\|x_n\|^{p-1}$ we obtain
$$A(v_n) + w_n = t_n \frac{u_n}{\|x_n\|^{p-1}} + (1 - t_n) K(v_n) \text{ for all } n \geq 1,$$
where $w_n \in \partial \psi_{t_n \|x_n\|}(v_n)$ and $u_n \in N(x_n)$. Evidently $\{A(v_n) + w_n\}_{n \geq 1} \subseteq L^q(Z)$ is bounded and so we may assume that
$$A(v_n) + w_n \xrightarrow{w} f \text{ in } L^q(Z) \text{ as } n \to \infty$$
and
$$\lim_{n \to \infty} \langle A(v_n) + w_n, v_n - v \rangle$$
$$= \lim_{n \to \infty} \int_Z \left(t_n \frac{u_n}{\|x_n\|^{p-1}} + (1 - t_n) K(v_n) \right) (v_n - v)$$
$$= 0$$

Recall that $(t, x) \to A(x) + \partial \psi_t(x)$ is a pseudomonotone homotopy (see the proof of Proposition 14). Hence it follows that
$$f = A(v)$$
(recall that $t_n \|x_n\| \to 0$ and $C(0) = W_0^{1,p}(Z)$) and
$$\langle A(v_n) + w_n, v_n \rangle \to \langle A(v), v \rangle.$$
Then exploiting the monotonicity of the convex subdifferential, we have
$$\langle w_n, v_n - v \rangle \geq 0,$$
hence
$$\langle A(v_n), v_n - v \rangle = \langle A(v_n) + w_n, v_n - v \rangle - \langle w_n, v_n - v \rangle$$
$$\leq \langle A(v_n) + w_n, v_n - v \rangle$$
from which it follows that
$$\limsup_{n \to \infty} \langle A(v_n), v_n - v \rangle \leq 0.$$
Since A is of type $(S)_+$, we have $v_n \to v$ in $W_0^{1,p}(Z)$ and so $\|v\| = 1$, therefore
$$v \neq 0.$$
Arguing as in the proof of Proposition 14 (keeping the notation introduced there), in the limit we obtain
$$A(v) = \widehat{g}_t K(v),$$
hence
$$v = 0$$
(since $\eta(z) \leq \widehat{g}_t(z) \leq \widehat{\eta}(z)$ a.e. on Z), a contradiction. So, from the homotopy invariance of the degree, we have
$$\deg(\partial \widehat{\varphi} + \partial \psi, B_\rho(0), 0) = \deg_{(S)_+}(A - hK, B_\rho(0), 0) = -1$$
for all $0 < \rho \leq \rho_0$, where the last equality follows from Drabek-Kufner-Nicolosi [15] (Chapter 3.6). $\square$

Also, as before, we can show that the functional $\varphi = \widehat{\varphi} + \psi$ admits a global minimizer.

PROPOSITION 24. *If hypotheses $(H_j)^3$ hold, then we can find $x_0 \in W_0^{1,p}(Z)$ such that*
$$\varphi(x_0) = \inf_{x \in W_0^{1,p}(Z)} \varphi(x).$$

Finally, Propositions 22, 23 and 24 and the additivity of the domain property of the degree lead to the following multiplicity result.

THEOREM 25. *If hypotheses $(H_j)^3$ hold, then problem (4.46) has at least two nontrivial solutions $x_0, \widehat{x} \in W_0^{1,p}(Z)$.*

CHAPTER 5

Hemivariational Inequalities with an Asymmetric Subdifferential

As before let $Z \subseteq \mathbb{R}^N$ be a bounded domain with a C^2–boundary ∂Z. In this Chapter we study the following nonlinear elliptic equation with a nonsmooth potential (hemivariational inequality):

(5.1) $$\begin{cases} -div\left(\|Dx(z)\|_{\mathbb{R}^N}^{p-2} Dx(z)\right) \in \partial j(z, x(z)) \text{ a.e. on } Z \\ x(z) = 0 \text{ on } \partial Z, \ 1 < p < \infty \end{cases}$$

In this case the assumptions on the nonsmooth potential $j(z, x)$ are:

$(H_j)^4$ $j : Z \times \mathbb{R}$ is a function such that that $j(z, 0) = 0$ a.e. on Z and
  (i) for all $x \in \mathbb{R}$, $z \to j(z, x)$ is measurable;
  (ii) for almost all $z \in Z$, $x \to j(z, x)$ is locally Lipschitz;
  (iii) for almost all $z \in Z$, all $x \in \mathbb{R}$ and all $u \in \partial j(z, x)$, we have

$$|u| \leq a(z) + c|x|^{p-1} \text{ with } a \in L^\infty(Z)_+, \ c > 0;$$

for almost all $z \in Z$, all $x \geq 0$ and all $u \in \partial j(z, x)$, we have

$$-c_0 x^{p-1} \leq u \text{ with } c_0 > 0;$$

and

$$\partial j(z, 0) \subseteq \mathbb{R}_+ \text{ a.e. on } Z;$$

  (iv) there exists $\theta \in L^\infty(Z)_+$ such that $\theta(z) \leq \lambda_1$ a.e. on Z with strict inequality on a set of positive measure, and

$$\limsup_{x \to +\infty} \frac{u}{x^{p-1}} \leq \theta(z)$$

uniformly for almost all $z \in Z$ and all $u \in \partial j(z, x)$;

  (v) there exist functions β and $\widehat{\beta} \in L^\infty(Z)_+$ such that $\beta(z) \geq \lambda_1$ a.e. on Z with strict inequality on a set of positive measure and

$$\beta(z) \leq \liminf_{x \to -\infty} \frac{u}{|x|^{p-2} x} \leq \limsup_{x \to -\infty} \frac{u}{|x|^{p-2} x} \leq \widehat{\beta}(z)$$

uniformly for almost all $z \in Z$ and all $u \in \partial j(z, x)$;

  (vi) there exist functions $\eta, \widehat{\eta} \in L^\infty(Z)_+$ such that $\eta(z) \geq \lambda_1$ a.e. on Z with strict inequality on a set of positive measure and

$$\eta(z) \leq \liminf_{x \to 0+} \frac{u}{x^{p-1}} \leq \limsup_{x \to 0+} \frac{u}{x^{p-1}} \leq \widehat{\eta}(z)$$

uniformly for almost all $z \in Z$ and all $u \in \partial j(z, x)$ and also we have

$$\lim_{x \to 0-} \frac{u}{|x|^{p-2} x} = 0$$

uniformly for almost all $z \in Z$ and all $u \in \partial j(z,x)$.

REMARK 26. *Hypotheses $(H_j)^4$ (iv) and (v) are nonuniform nonresonance conditions at $+\infty$ and $-\infty$ respectively. Note that according to $(H_j)^4$ (iv) near $+\infty$, the slopes*

$$\left\{\frac{u}{|x|^{p-2}x}\right\}_{u \in \partial j(z,x)}$$

stay below the principal eigenvalue $\lambda_1 > 0$, while near $-\infty$ they stay above $\lambda_1 > 0$. This is an interesting situation, which creates additional difficulties due to the fact that it is not clear in this setting how to produce linking sets. So the variational methods need further refinement in order to work; see Fan-Huang-Zhou [16] *and Kyritsi-Papageorgiou* [29], [30]. *Hypothesis $(H_j)^4$ (vi) is a nonuniform nonresonance condition at zero from the right and a nonresonance condition at zero from the left. We point out that the asymmetric behavior is also present near zero. We would also like to mention that $(H_j)^4$ (iv) implies that*

$$(5.2) \qquad \limsup_{x \to +\infty} \frac{pj(z,x)}{x^p} \leq \theta(z)$$

uniformly for almost all $z \in Z$.

Indeed, because of hypothesis $(H_j)^4$ (ii) and Rademacher's theorem, we know that for almost all $z \in Z$ and almost all $x \in \mathbb{R}$, $\frac{d}{dx}j(z,x)$ exists and

$$(5.3) \qquad \frac{d}{dx}j(z,x) \in \partial j(z,x).$$

By virtue of hypothesis $(H_j)^4$ (iv) given $\varepsilon > 0$, we can find $M = M(\varepsilon) > 0$ such that

$$(5.4) \qquad u \leq (\theta(z) + \varepsilon) x^{p-1}$$

for almost all $z \in Z$, all $x \geq M$ and all $u \in \partial j(z,x)$. So

$$j(z,x) = j(z,M) + \int_M^x \frac{d}{dr}j(z,r)\,dr$$
$$\leq j(z,M) + \int_M^x (\theta(z) + \varepsilon)r^{p-1}\,dr \quad (see\ (5.3)\ and\ (5.4))$$
$$= j(z,M) + \frac{1}{p}(\theta(z) + \varepsilon)(x^p - M^p) \text{ for a.a. } z \in Z \text{ and all } x \geq M,$$

hence

$$(5.5) \qquad \frac{pj(z,x)}{x^p} \leq \frac{pj(z,M)}{x^p} + (\theta(z) + \varepsilon)\left(1 - \frac{M^p}{x^p}\right)$$

From hypothesis $(H_j)^4$ (iii) and the mean value theorem for locally Lipschitz functions, we have

$$(5.6) \qquad |j(z,x)| \leq \widehat{a}(z) + \widehat{c}|x|^p$$

for almost all $z \in Z$ and all $x \in \mathbb{R}$, with $\widehat{a} \in L^\infty(Z)_+$ and $\widehat{c} > 0$.

Fix $\varepsilon > 0$. Because of (5.6) we can find $\overline{M} = \overline{M}(\varepsilon) \geq M > 0$ such that for almost all $z \in Z$ and all $x \geq \overline{M}$, we have

(5.7) $$\frac{p|j(z, M)|}{x^p} \leq \varepsilon.$$

Using (5.7) in (5.5), we see that

$$\frac{pj(z, x)}{x^p} \leq \theta(z) + 2\varepsilon,$$

for almost all $z \in Z$ and all $x \geq \overline{M}$. From this we infer that (5.2) holds.

A nonsmooth potential function satisfying hypotheses $(H_j)^4$ is given below. For simplicity we drop the z-dependence:

$$j(x) = \begin{cases} \frac{\mu}{p}|x|^p + \frac{1}{s} - \frac{\mu}{p} & \text{if } x < -1 \\ \frac{1}{s}|x|^s & \text{if } x \in [-1, 0] \\ \frac{\xi-1}{p}x^p + \frac{1}{p}\sin x^p & \text{if } x \in [0, 1] \\ \frac{\theta}{p}x^p + \frac{\xi-1}{p} + \frac{\sin 1}{p} - \frac{\theta}{p} & \text{if } x > 1 \end{cases}$$

with $\theta < \lambda_1 < \xi$, $\mu, p < s$.

The Euler functional $\varphi : W_0^{1,p}(Z) \to \mathbb{R}$ corresponding to problem (5.1) is given by

$$\varphi(x) = \frac{1}{p}\|Dx\|_p^p - \int_Z j(z, x(z))\, dz \text{ for all } x \in W_0^{1,p}(Z).$$

We know (see for example Gasinski-Papageorgiou [**17**], p.58) that φ is Lipschitz continuous on bounded sets, hence locally Lipschitz. Also we introduce the following Banach space

$$C_0^1(\overline{Z}) := \{x \in C^1(\overline{Z}) : x\,|_{\partial Z} = 0\}$$

equipped with the supremum norm.

Using the usual pointwise ordering (i.e., $x \leq y$ in $C_0^1(\overline{Z})$ if and only if $x(z) \leq y(z)$ for all $z \in Z$), $C_0^1(\overline{Z})$ becomes an ordered Banach space, with the order cone given by

$$C_0^1(\overline{Z})_+ := \{x \in C_0^1(\overline{Z}) : x(z) \geq 0 \text{ for all } z \in \overline{Z}\}.$$

This closed, convex cone (order cone) has a nonempty interior given by

$$\text{int } C_0^1(\overline{Z})_+ := \left\{x \in C_0^1(\overline{Z}) : x(z) > 0\, \forall z \in Z \text{ and } \frac{\partial x}{\partial n}(z) < 0\, \forall z \in \partial Z\right\}.$$

Here $n(z)$ denotes the unit outward normal vector at $z \in \partial Z$.

Next we will produce a first nontrivial solution for problem (5.1) by minimizing a suitable modification of the Euler functional φ.

To do this we will need a result (see Theorem 27 below) which relates local minimizers in $C_0^1(\overline{Z})$ and in $W_0^{1,p}(Z)$ respectively, for a large class of nonsmooth functionals. This way we are able to link results based on the strong maximum principle (linear and nonlinear formulations of it) and results coming from the use of variational arguments.

This result for $p = 2$ (semilinear case) and for a smooth potential, was first proved by Brezis-Nirenberg [**7**]. It was extended to nonlinear nonsmooth problems by Kyritsi - Papageorgiou [**29**] (see also Gasinski-Papageorgiou [**18**], p.655).

Before stating the result let us state the hypotheses on the nonsmooth potential, which are very general

$\left(H_{\widehat{j}}\right)$ $\widehat{j}: Z \times \mathbb{R} \to \mathbb{R}$ is a function such that that

(i) for every $x \in \mathbb{R}$, $z \to \widehat{j}(z,x)$ is measurable and $\widehat{j}(.,0) \in L^\infty(Z)$;
(ii) for almost all $z \in Z$, $x \to \widehat{j}(z,x)$ is locally Lipschitz;
(iii) for almost all $z \in Z$, all $x \in \mathbb{R}$ and all $u \in \partial j(z,x)$, we have

$$|u| \leq \widehat{a}(z) + \widehat{c}|x|^{s-1}$$

with $a \in L^\infty(Z)_+$, $\widehat{c} > 0$ and $1 \leq s < p^*$, where

$$p^* = \begin{cases} \frac{Np}{N-p} & \text{if } p < N \\ +\infty & \text{if } p \geq N \end{cases}.$$

We consider the functional $\psi : W_0^{1,p}(Z) \to \mathbb{R}$ defined by

$$\psi(x) = \frac{1}{p}\|Dx\|_p^p - \int_Z \widehat{j}(z, x(z))\,dz \text{ for all } x \in W_0^{1,p}(Z).$$

We know that ψ is locally Lipschitz.

THEOREM 27. *If hypotheses $\left(H_{\widehat{j}}\right)$ hold and $x_0 \in W_0^{1,p}(Z)$ is a local $C_0^1(\overline{Z})$ – minimizer of ψ i.e., there exists $r > 0$ such that*

$$\psi(x_0) \leq \psi(x_0 + y) \text{ for all } y \in C_0^1(\overline{Z}) \text{ with } \|y\|_{C_0^1(\overline{Z})} \leq r,$$

then x_0 is also a minimizer of ψ, in $W_0^{1,p}(Z)$, i.e., there exists $r_0 > 0$ such that

$$\psi(x_0) \leq \psi(x_0 + y) \text{ for all } y \in W_0^{1,p}(Z) \text{ with } \|y\| \leq r_0.$$

We will use this theorem to produce a nontrivial positive solution for problem (5.1).

PROPOSITION 28. *If hypotheses $(H_j)^4$ (i), (ii), (iii), (iv) and (vi) hold, then problem (5.1) has a solution $x_0 \in \text{int } C_0^1(\overline{Z})_+$ which is a local minimizer of the Euler funcional φ.*

PROOF. Let $\tau_+ : \mathbb{R} \to \mathbb{R}_+$ be the truncation map defined by

$$\tau_+(x) = \begin{cases} 0 & \text{if } x \leq 0 \\ x & \text{if } x > 0 \end{cases}.$$

Clearly τ_+ is Lipschitz continuous. So if we set

$$j_+(z,x) = j(z, \tau_+(x)),$$

then we see that:

(i)' for all $x \in \mathbb{R}$, $z \to j_+(z,x)$ is measurable;
(ii)' for almost all $z \in Z$, $x \to j_+(z,x)$ is locally Lipschitz.

Also, by $(H_j)^4$ (iii), it is easily verified that $j_+(.,.)$ satisfies condition (iii) in $\left(H_{\widehat{j}}\right)$ with $s = p$. We consider the functional $\varphi_+ : W_0^{1,p}(Z) \to \mathbb{R}$ defined by

$$\varphi_+(z) = \frac{1}{p}\|Dx\|_p^p - \int_Z j_+(z, x(z))\,dz \text{ for all } x \in W_0^{1,p}(Z).$$

We know that φ_+ is Lipschitz continuous on bounded sets, hence locally Lipschitz. Moreover, exploiting the compact embedding of $W_0^{1,p}(Z)$ into $L^p(Z)$ and the weak lower semicontinuity of the norm functional in a Banach space, we can easily check that φ_+ is weakly lower semicontinuous on $W_0^{1,p}(Z)$. Because of the hypothesis $(H_j)^4\,(iv)$, given $\varepsilon > 0$, we can find $M_1 = M_1(\varepsilon) > 0$ such that for almost all $z \in Z$, all $x \geq M$ and all $u \in \partial j(z,x)$, we have

(5.8) $$u \leq (\theta(z) + \varepsilon) x^{p-1}$$

From the nonsmooth chain rule (see Clarke [**10**], p.42 and Gasinski-Papageorgiou [**18**], p.55), we have

(5.9) $$\partial j_+(z,x) \subseteq \begin{cases} \{0\} & \text{if } x < 0 \\ \{\lambda \partial j(z,0)\}_{\lambda \in [0,1]} & \text{if } x = 0 \\ \partial j(z,x) & \text{if } x > 0 \end{cases}.$$

Combining (5.8) and (5.9) and noting that

$$\frac{d}{dx} j_+(z,x) \in \partial j_+(z,x),$$

we obtain

(5.10) $$j_+(z,x) \leq j_+(z,M_1) + \frac{1}{p}(\theta(z) + \varepsilon) x^p, \text{ for a.a. } z \in Z \text{ and all } x \geq M_1.$$

On the other hand, using $(H_j)^4\,(iii)$, the mean value theorem for locally Lipschitz functions (see Clarke [**10**], p.41 and Gasinski-Papageorgiou [**18**], p.53) and recalling that $j(z,0) = 0$ a.e. on Z, we see that for almost all $z \in Z$ and all $x < M_1$, we have

(5.11) $$j_+(z,x) \leq \beta_\varepsilon(z) \text{ with } \beta_\varepsilon \in L^\infty(Z)_+.$$

Taking into account (5.10) and (5.11) and the fact that $j_+(.,M_1) \in L^\infty(Z)$ (see $(H_j)^4\,(iii)$), we obtain

(5.12) $$j_+(z,x) \leq \frac{1}{p}(\theta(z) + \varepsilon)|x|^p + \widehat{\beta}_\varepsilon(z) \text{ with } \widehat{\beta}_\varepsilon \in L^\infty(Z)_+.$$

So for every $x \in W_0^{1,p}(Z)$, we have

$$\varphi_+(x) = \frac{1}{p}\|Dx\|_p^p - \int_Z j_+(z, x(z))\,dz$$

$$\geq \frac{1}{p}\|Dx\|_p^p - \frac{1}{p}\int_Z (\theta(z))|x(z)|^p\,dz - \frac{\varepsilon}{p}\|x\|_p^p - \|\widehat{\beta}_\varepsilon\|_1 \text{ (see (5.12))}$$

(5.13) $$\geq \frac{1}{p}\left(\xi - \frac{\varepsilon}{\lambda_1}\right)\|Dx\|_p^p - \|\widehat{\beta}_\varepsilon\|_1 \text{ (see Lemma 12 and (2.1) with } m \equiv 1)$$

Choosing $\varepsilon < \lambda_1 \xi$, from (5.13) and Poincare's inequality, we infer that φ_+ is coercive. This fact, together with the weak lower semicontinuity of φ_+, via the Weierstrass theorem, implies that we can find $x_0 \in W_0^{1,p}(Z)$ such that

$$\varphi_+(x_0) = \min\{\varphi_+(x) : x \in W_0^{1,p}(Z)\},$$

hence

$$0 \in \partial \varphi_+(x_0),$$

and therefore

(5.14) $$A(x_0) = u_0,$$

with
$$u_0 \in L^q(Z), u_0(z) \in \partial j_+(z, x_0(z)) \text{ a.e. on } Z.$$
In (5.14) we act with the test function $-x_0^- \in W_0^{1,p}(Z)$. So we obtain
$$\|Dx_0^-\|_p^p = -\int_Z u_0 x_0^- dz.$$
Recall that
$$Dx_0^-(z) = \begin{cases} -Dx_0(z) & \text{if } x_0(z) < 0, \\ 0 & \text{otherwise.} \end{cases}$$
But from (5.9) we deduce that
$$u_0(z) = 0 \text{ a.e. on } \{x_0 < 0\}.$$
So it follows that
$$\int_Z u_0(z) x_0^-(z) dz = 0,$$
hence
$$\|Dx_0^-\|_p = 0,$$
i.e., $x_0^- = 0$ and so
$$x_0 \geq 0.$$
Next we have to show that $x_0 \neq 0$. Because of hypothesis $(H_j)^4 (vi)$, given $\varepsilon > 0$ we can find $\delta = \delta(\varepsilon) > 0$ such that
$$(\eta(z) - \varepsilon) x^{p-1} \leq u \text{ for a.a. } z \in Z, \text{ all } 0 \leq x < \delta \text{ and all } u \in \partial j(z,x),$$
hence
$$(\eta(z) - \varepsilon) x^{p-1} \leq \frac{d}{dx} j(z,x) \text{ for a.a. } z \in Z, \forall x \in [0,\delta] \setminus D(z) \text{ with } |Dz|_1 = 0,$$
where by $|\cdot|_1$ we denote the Lebesgue measure on $\mathbb{R}$. Integrating this inequality, we obtain
$$(5.15) \quad \frac{1}{p} (\eta(z) - \varepsilon) x^p \leq j_+(z,x) = j(z,x) \text{ for a.a. } z \in Z \text{ and all } x \in [0,\delta].$$
Recall that $u_1 \in \text{int} C_0^1(\overline{Z})_+$, where u_1 is the eigenfunction corresponding to λ_1 (see Chapter 2). So we can find $t_0 > 0$ such that for all $t \in (0, t_0]$, we have
$$(5.16) \quad 0 < tu_1(z) \leq \delta \text{ for all } z \in Z.$$
From (5.15) and (5.16) it follows that
$$(5.17) \quad j_+(z, tu_1(z)) \geq \frac{t^p}{p} (\eta(z) - \varepsilon) u_1(z)^p \text{ for a.a. } z \in Z, \forall 0 < t \leq t_0.$$
So we obtain
$$\varphi_+(tu_1) = \frac{t^p}{p} \|Du_1\|_p^p - \int_Z j_+(z, tu_1(z)) dz$$
$$\leq \frac{t^p}{p} \|Du_1\|_p^p - \frac{t^p}{p} \int_Z \eta(z) u_1(z)^p dz + \frac{t^p}{p} \varepsilon \|u_1\|_p^p \text{ (see (5.17))}$$
$$= \frac{t^p}{p} \int_Z (\lambda_1 - \eta(z)) u_1(z)^p dz + \frac{t^p}{p} \varepsilon \|u_1\|_p^p$$
By virtue of the hypothesis on η and since $u_1(z) > 0$ for all $z \in Z$, we see that
$$-\int_Z (\lambda_1 - \eta(z)) u_1(z)^p dz = \gamma > 0.$$

So if we choose $\varepsilon > 0$ small, then
$$\varphi_+(tu_1) \leq \frac{t^p}{p}\left(-\gamma + \varepsilon \|u_1\|_p^p\right) < 0 \text{ for all } t \in (0, t_0],$$
hence
$$\varphi_+(x_0) = \inf\left\{\varphi_+(x) : x \in W_0^{1,p}(Z)\right\} < \varphi_+(0) = 0,$$
therefore $x_0 \neq 0$. From (5.14) we deduce that

(5.18) $$\begin{cases} -div\left(\|Dx_0(z)\|_{\mathbb{R}^N}^{p-2} Dx_0(z)\right) = u_0(z) \text{ a.e. on } Z \\ x_0 \equiv 0 \text{ on } \partial Z. \end{cases}$$

From (5.18) and the nonlinear regularity theory (see Gasinski-Papageorgiou [18], pp.115-116), we have that $x_0 \in C_0^1(\overline{Z})_+ \setminus \{0\}$. Note that $u_0(z) \in \partial j(z, x_0(z))$ a.e. on Z (see (5.9)) and so from hypothesis $(H_j)^4$ (iii) and (5.18) it follows that
$$\triangle_p x_0(z) = -u_0(z) \leq c_0 x_0(z)^{p-1} \text{ a.e. on } Z.$$
Invoking the nonlinear strong maximum principle of Vazquez [45] (see also Gasinski-Papageorgiou [18], pp.116-117), we obtain that
$$x_0(z) > 0 \text{ for all } z \in Z \text{ and } \frac{\partial x_0}{\partial n}(z) < 0 \text{ for all } x \in \partial Z,$$
hence $x_0 \in int\, C_0^1(\overline{Z})_+$. Therefore we can find $r_0 > 0$ such that
$$B_{r_0}^{C_0^1(\overline{Z})}(x_0) = \left\{x \in C_0^1(\overline{Z}) : \|x - x_0\|_{C_0^1(\overline{Z})} < r_0\right\} \subseteq C_0^1(\overline{Z})_+.$$
Then from the definition of the functional φ_+, we have
$$\varphi_+(x_0) = \varphi(x_0) \leq \varphi_+(y) = \varphi(y) \text{ for all } y \in B_{r_0}^{C_0^1(\overline{Z})}(x_0),$$
which implies that x_0 is a local $C_0^1(\overline{Z})$–minimizer of φ. Because of Theorem 27, it follows that x_0 is a local $W_0^{1,p}(Z)$-minimizer of φ as claimed in the proposition. □

We may assume that $x_0 \in int C_0^1(\overline{Z})_+$ is an isolated critical point of φ, or otherwise we will have a continuum of nontrivial critical points of φ, hence a continuum of nontrivial solutions of (5.1).

PROPOSITION 29. *If hypotheses $(H_j)^4$ (i), (ii), (iii), (iv) and (vi) hold and $x_0 \in int\, C_0^1(\overline{Z})_+$ is as in Proposition 28, then we can find $r > 0$ such that*
$$\deg(\partial\varphi, B_r(x_0), 0) = 1$$

PROOF. Since x_0 is an isolated critical point of φ, we can find $r_0 > 0$ such that

(5.19) $\quad \varphi(x_0) < \varphi(y)$ and $0 \notin \partial\varphi(y)$ for all $y \in \overline{B}_{r_0}(x_0) \setminus \{x_0\}$

where $\overline{B}_{r_0}(x_0) = \left\{x \in W_0^{1,p}(Z) : \|x - x_0\| \leq r_0\right\}$. **Claim:** For all $0 < r < r_0$ we have

(5.20) $$\inf\left\{\varphi(x) : x \in \overline{B}_{r_0}(x_0) \setminus B_r(x_0)\right\} > \varphi(x_0).$$

We argue indirectly. So suppose that the Claim is not true. Then we can find $r \in (0, r_0)$ and a sequence $\{x_n\}_{n \geq 1} \subseteq \overline{B}_{r_0}(x_0) \setminus B_{r_0}(x_0)$ such that
$$\varphi(x_n) \downarrow \varphi(x_0) \text{ as } n \to \infty.$$

The sequence $\{x_n\}_{n\geq 1} \subseteq W_0^{1,p}(Z)$ is bounded. So by passing to a suitable subsequence if necessary, we may assume that

$$x_n \xrightarrow{w} \widehat{x} \text{ in } W_0^{1,p}(Z) \text{ and } x_n \to \widehat{x} \text{ in } L^p(Z).$$

(Recall that $W_0^{1,p}(Z)$ is embedded compactly in $L^p(Z)$). Exploiting the weak lower semicontinuity of the Euler functional φ, we have

(5.21) $$\varphi(\widehat{x}) \leq \lim_{n\to\infty} \varphi(x_n) = \varphi(x_0).$$

Evidently, by (5.19), $\widehat{x} \in \overline{B}_{r_0}(x_0)$ and so we must have

(5.22) $$\varphi(x_0) \leq \varphi(\widehat{x}) \text{ with strict inequality if } x_0 \neq \widehat{x}.$$

From (5.21) and (5.22) it follows that $x_0 = \widehat{x}$. Invoking the mean value theorem for locally Lipschitz functions, we can find

$$w_n^* \in \partial\varphi\left(\lambda_n x_n + (1-\lambda_n)\frac{x_n + x_0}{2}\right) \text{ with } 0 < \lambda_n < 1, \ n \geq 1,$$

such that

$$\varphi(x_n) - \varphi\left(\frac{x_n + x_0}{2}\right) = \left\langle w_n^*, \frac{x_n - x_0}{2}\right\rangle,$$

where as before by $\langle \cdot, \cdot \rangle$ we denote the duality brackets for the pair

$$\left(W^{-1,q}(Z), W_0^{1,p}(Z)\right).$$

We know that

$$w_n^* = A\left(\lambda_n x_n + (1-\lambda_n)\frac{x_n + x_0}{2}\right) - u_n$$

with $u_n \in L^q(Z)$,

$$u_n(z) \in \partial j\left(z, \lambda_n x_n(z) + (1-\lambda_n)\frac{x_n(z) + x_0(z)}{2}\right) \text{ a.e. on } Z, n \geq 1.$$

Hence

(5.23) $$\varphi(x_n) - \varphi\left(\frac{x_n + x_0}{2}\right) = \frac{1}{2}\left\langle A\left(\lambda_n x_n + (1-\lambda_n)\frac{x_n + x_0}{2}\right), x_n - x_0\right\rangle \\ - \frac{1}{2}\int_Z u_n(z)(x_n(z) - x_0(z))\,dz.$$

By hypothesis $(H_j)^4$ (iii) the sequence $\{u_n\}_{n\geq 1} \subseteq L^q(Z)$ is bounded. Also recall that $x_n \to x_0$ in $L^p(Z)$. So it follows that

(5.24) $$\int_Z u_n(z)(x_n(z) - x_0(z))\,dz \to 0 \text{ as } n \to \infty.$$

Also from the choice of the sequence $\{x_n\}_{n\geq 1} \subseteq W_0^{1,p}(Z)$, we have

(5.25) $$\varphi(x_n) \to \varphi(x_0) \text{ as } n \to \infty.$$

Moreover, since φ is weakly lower semicontinuous and

$$\frac{x_n + x_0}{2} \xrightarrow{w} x_0 \text{ in } W_0^{1,p}(Z),$$

we have

(5.26) $$\varphi(x_0) \leq \liminf_{n\to\infty} \varphi\left(\frac{x_n + x_0}{2}\right).$$

Therefore, if in (5.23) we pass to the limit as $n \to \infty$ and use (5.24) (5.25) and (5.26), we obtain

(5.27) $$\limsup_{n\to\infty} \left\langle A\left(\lambda_n x_n + (1-\lambda_n)\frac{x_n+x_0}{2}\right), x_n - x_0 \right\rangle \leq 0$$

Clearly we may assume that $\lambda_n \to \lambda \in [0,1]$ as $n \to \infty$. Since A is an operator of type $(S)_+$ (see Lemma 11) and

$$\lambda_n x_n + (1-\lambda_n)\frac{x_n+x_0}{2} \xrightarrow{w} x_0 \text{ in } W_0^{1,p}(Z),$$

from (5.27) we deduce that

(5.28) $$\lambda_n x_n + (1-\lambda_n)\frac{x_n+x_0}{2} \to x_0 \text{ in } W_0^{1,p}(Z).$$

Note that

(5.29) $$\left\|\lambda_n x_n + (1-\lambda_n)\frac{x_n+x_0}{2} - x_0\right\| = (1+\lambda_n)\left\|\frac{x_n-x_0}{2}\right\| \geq \frac{r}{2}.$$

Comparing (5.28) and (5.29), we reach a contradiction. This proves the Claim. Next we set

(5.30) $$\mu = \inf\left\{\varphi(x) : x \in B_{r_0}(x_0) \setminus B_{\frac{r_0}{2}}(x_0)\right\} - \varphi(x_0).$$

Because of (5.20) we have $\mu > 0$. We introduce the set

(5.31) $$V = \left\{x \in B_{\frac{r_0}{2}}(x_0) : \varphi(x) - \varphi(x_0) < \mu\right\}.$$

Obviously, $x_0 \in V$, hence V is nonempty and open (since φ is continuous). So we can find $0 < r < \frac{r_0}{2}$ such that $\overline{B}_r(x_0) \subseteq V$. Therefore we can apply Proposition 5 with data $U = B_{r_0}(x_0)$, $\widehat{\varphi} = \varphi - \varphi(x_0)$, x_0, $\mu > 0$ as above and

(5.32) $$0 < \xi < \inf\{\varphi(x) : x \in B_{r_0}(x_0) \setminus B_r(x_0)\} - \varphi(x_0) \text{ (see (5.20))}.$$

Indeed, note that since $r < \frac{r_0}{2}$, from (5.30) and (5.31), we have $\overline{V} \subseteq B_{r_0}(x_0)$. Moreover, from (5.31) and (5.32), we have

$$\{x \in B_{r_0}(x_0) : \varphi(x) - \varphi(x_0) \leq \xi\} \subseteq B_r(x_0) \subseteq \overline{B}_r(x_0) \subseteq V.$$

Recall that $0 \notin \partial\varphi(x)$ for all $x \in \overline{B}_{r_0}(x_0) \setminus \{x_0\}$ (see (5.19)), hence $0 \notin \partial\varphi(x)$ for all $x \in \overline{B}_{r_0}(x_0)$ with $\xi \leq \varphi(x) - \varphi(x_0) \leq \mu$. So applying Proposition 5, we have

$$\deg(\partial\varphi, V, 0) = 1.$$

On the other hand, from the previous arguments we know that

$$0 \notin \partial\varphi\left(\overline{V} \setminus B_r(x_0)\right).$$

So the excision property of the degree map $\deg_{(S)_+}$, implies that

$$\deg(\partial\varphi, B_r(x_0), 0) = 1.$$

□

Next we evaluate the degree of $\partial\varphi$ in large balls.

PROPOSITION 30. *If hypotheses* $(H_j)^4$ *hold, then there exists* $R_0 > 0$ *such that*
$$\deg(\partial\varphi, B_R(0), 0) = 0 \text{ for all } R \geq R_0.$$

PROOF. Recall that $N : W_0^{1,p}(Z) \to 2^{L^q(Z)} \setminus \{\varnothing\}$ is defined by
$$N(x) = \{u \in L^q(Z) : u(z) \in \partial j(z, x(z)) \text{ a.e. on } Z\}.$$
From Proposition 3, we know that N is usc from $W_0^{1,p}(Z)$ with the norm topology into $L^q(Z)$ with the weak topology and has weakly compact and convex values. Moreover, for every $x \in W_0^{1,p}(Z)$, we have
$$\partial \varphi(x) = A(x) - N(x).$$
Let $K_- : L^p(Z) \to L^q(Z)$ be the strictly monotone, continuous (hence maximal monotone too) operator defined by
$$K_-(x)(.) = |x^-(.)|^{p-2} x^-(.)$$
(Recall that for every $x \in W_0^{1,p}(Z)$, $x^- = \max\{-x, 0\}$ (the negative part of x) and $x^+ = \max\{x, 0\}$ (the positive part of x); we know that $x^+, x^- \in W_0^{1,p}(Z)$. Since $L^q(Z)$ is compactly embedded into $W^{-1,q}(Z) = W_0^{1,p}(Z)^*$ (Schauder's theorem), it follows that K_- considered as a map from $W_0^{1,p}(Z)$ into $W^{-1,q}(Z)$ is completely continuous (hence compact too; see Gasinski-Papageorgiou [18], p.68). Also, let $h \in L^\infty(Z)_-$ be such that
$$-\widehat{\beta}(z) \le h(z) \le -\beta(z) \text{ a.e. on } Z \text{ (see hypothesis } (H_j)^4 (v)).$$
We consider the admissible homotopy $h_1 : [0,1] \times W_0^{1,p}(Z) \to 2^{W^{-1,q}(Z)} \setminus \{\varnothing\}$ defined by
$$h_1(t, x) = Ax - tN(x) - (1-t) h K_-(x).$$
Claim: There exists $R_0 > 0$ such that for all $R \ge R_0$ and all $t \in [0,1]$ we have
(5.33) $\quad 0 \ne Ax - tN(x) - (1-t) h K_-(x)$ for all $x \in \partial B_R(0)$.

Again we argue by contradiction. So suppose that the Claim is not true. Then we can find $\{x_n\}_{n \ge 1} \subseteq W_0^{1,p}(Z)$ and $\{t_n\}_{n \ge 1} \subseteq [0,1]$ such that
$$\|x_n\| \to \infty, \ t_n \to t \in [0,1] \text{ as } n \to \infty$$
and
$$Ax_n - t_n N(x_n) - (1 - t_n) h K_-(x_n) = 0 \text{ with } u_n \in N(x_n) \text{ for all } n \ge 1.$$
On the last equation we act with the test function $x_n^+ \in W_0^{1,p}(Z)$ and we obtain
(5.34) $\quad \|Dx_n^+\|_p^p = t_n \int_Z u_n(z) x_n^+(z) dz.$

Because of hypothesis $(H_j)^4 (iv)$, given $\varepsilon > 0$, we can find $M_2 = M_2(\varepsilon) > 0$ such that
$$u \le (\theta(z) + \varepsilon) x^{p-1} \text{ for a.a. } z \in Z, \text{ all } x \ge M \text{ and all } u \in \partial j(z, x).$$
On the other hand, hypothesis $(H_j)^4 (iii)$ and the mean value theorem for locally Lipschitz functions, yield a function $\gamma_\varepsilon \in L^\infty(Z)_+$ such that
$$u \le \gamma_\varepsilon(z) \text{ for a.a. } z \in Z, \text{ all } x \in [0, M] \text{ and all } u \in \partial j(z, x).$$
So finally we can write
$$u \le (\theta(z) + \varepsilon) x^{p-1} + \gamma_\varepsilon(z) \text{ for a.a. } z \in Z, \text{ all } x \ge 0 \text{ and all } u \in \partial j(z, x).$$
Hence, for all $n \ge 1$
(5.35) $\quad u_n(z) x_n^+(z) \le (\theta(z) + \varepsilon) x_n^+(z)^p + \gamma_\varepsilon(z) x_n^+(z)$ a.e. on Z.

Using (5.35) in (5.34), we obtain for all $n \geq 1$

$$\left\| Dx_n^+ \right\|_p^p - \int_Z \theta(z) x_n^+(z)^p \, dz - \varepsilon \left\| x_n^+ \right\|_p^p \leq c_\varepsilon \left\| x_n^+ \right\|_p \text{ for some } c_\varepsilon > 0,$$

hence

(5.36) $\quad \xi \left\| Dx_n^+ \right\|_p^p - \dfrac{\varepsilon}{\lambda_1} \left\| Dx_n^+ \right\|_p^p \leq c_\varepsilon' \left\| Dx_n^+ \right\|_p \text{ for some } c_\varepsilon' > 0.$

To obtain (5.36) we have used Lemma 12 and the variational characterization of λ_1 (see (2.1)). If we choose $\varepsilon < \lambda_1 \xi$, from (5.36) and Poincare's inequality we infer that

$$\{x_n^+\}_{n \geq 1} \text{ is bounded in } W_0^{1,p}(Z).$$

By hypothesis, $\lim_{n \to \infty} \|x_n\| = \infty$. Therefore

$$\|x_n^-\| \to \infty \text{ as } n \to \infty.$$

We set

$$y_n = \dfrac{x_n^-}{\|x_n^-\|}, \; n \geq 1.$$

By passing to a suitable subsequence if necessary, we may assume that

$$y_n \xrightarrow{w} y \text{ in } W_0^{1,p}(Z), \; y_n \to y \text{ in } L^p(Z), \; y_n(z) \to y(z) \text{ a.e. on } Z$$

and

$$y_n(z) \leq k(z) \text{ a.e. on } Z, \text{ for all } n \geq 1, \text{ with } k \in L^p(Z).$$

By hypothesis

$$A(x_n) = t_n u_n + (1 - t_n) h K_-(x_n).$$

Dividing with $\|x_n^-\|^{p-1}$ and noting that $A(x_n) = A(x_n^+) - A(x_n^-)$, we obtain

(5.37) $\quad \dfrac{1}{\|x_n^-\|^{p-1}} A(x_n^+) - A(y_n) = t_n \dfrac{u_n}{\|x_n^-\|^{p-1}} + (1 - t_n) h y_n^{p-1}.$

Acting with the test function $y_n - y \in W_0^{1,p}(Z)$ in (5.37) we obtain

$$\dfrac{1}{\|x_n^-\|^{p-1}} \langle A(x_n^+), y_n - y \rangle - \langle A(y_n), y_n - y \rangle$$

(5.38) $\quad = t_n \displaystyle\int_Z \dfrac{u_n}{\|x_n^-\|^{p-1}} (y_n - y) \, dz + (1 - t_n) \displaystyle\int_Z h y_n^{p-1} (y_n - y) \, dz.$

Since $\{x_n^+\}_{n \geq 1} \subseteq W_0^{1,p}(Z)$ is bounded and $\|x_n^-\| \to \infty$, we see that

$$\dfrac{1}{\|x_n^-\|^{p-1}} \langle A(x_n^+), y_n - y \rangle =$$

(5.39) $\quad \dfrac{1}{\|x_n^-\|^{p-1}} \displaystyle\int_Z \|Dx_n^+\|^{p-2} (Dx_n^+, Dy_n - Dy)_{\mathbb{R}^N} \, dz \to 0 \text{ as } n \to \infty.$

From hypothesis $(H_j)^4 \, (iii)$, we have

(5.40) $\quad \dfrac{|u_n(z)|}{\|x_n^-\|^{p-1}} \leq \dfrac{a(z)}{\|x_n^-\|^{p-1}} + c y_n(z)^{p-1} \leq k_0(z) \text{ a.e. on } Z$

for all $n \geq 1$, with $k_0 \in L^q(Z)$. Because of (5.40), we may assume that

(5.41) $\quad \dfrac{u_n}{\|x_n^-\|^{p-1}} \xrightarrow{w} f_0 \text{ in } L^q(Z) \text{ as } n \to \infty.$

Arguing as in the proof of Proposition 14, we can show that
$$-\widehat{\beta}(z) y(z)^{p-1} \leq f_0(z) \leq -\beta(z) y(z)^{p-1} \text{ a.e. on } \{y > 0\},$$
and
$$f_0(z) = 0 \text{ a.e. on } \{y = 0\} \text{ (see also (5.40))}.$$
Note that $y \geq 0$ and so $Z = \{y > 0\} \cup \{y = 0\}$. So it follows that
$$-\widehat{\beta}(z) y(z)^{p-1} \leq f_0(z) \leq \beta(z) y(z)^{p-1} \text{ a.e. on } Z,$$
hence
$$(5.42) \qquad f_0(z) = \widehat{\xi}(z) y(z)^{p-1} \text{ a.e. on } Z,$$
where $\widehat{\xi} \in L^\infty(Z)_+$ satisfies
$$-\widehat{\beta}(z) \leq \widehat{\xi}(z) \leq -\beta(z) \text{ a.e. on } Z.$$
So if we pass to the limit as $n \to \infty$ in (5.38) and we use (5.39) and (5.41), we obtain
$$\lim_{n \to \infty} \langle A(y_n), y_n - y \rangle = 0.$$
Since A is of type $(S)_+$ (see Lemma 11), we have
$$(5.43) \qquad y_n \to y \text{ in } W_0^{1,p}(Z).$$
Therefore if we pass to the limit as $n \to \infty$ in (5.37), we obtain
$$-A(y) = t\widehat{\xi} y^{p-1} + (1-t) h y^{p-1}$$
hence
$$(5.44) \qquad \begin{cases} -\text{div}\left(\|Dy(z)\|_{\mathbb{R}^N}^{p-2} Dy(z)\right) \\ = -\left(t\widehat{\xi}(z) + (1-t) h(z)\right) |y(z)|^{p-2} y(z) \text{ a.e. on } Z \\ y \equiv 0 \text{ on } \partial Z. \end{cases}$$
Note that because $\|y_n\| = 1$ for all $n \geq 1$ and $y_n \to y$ in $W_0^{1,p}(Z)$, we obtain $y \neq 0$. So from (5.44) we have that $y \in W_0^{1,p}(Z)_+$ is an eigenfunction of $\left(-\triangle_p, W_0^{1,p}(Z)\right)$ with weight $m = -\left[t\widehat{\xi} + (1-t) h\right] \in L^\infty(Z)_+$. Note that
$$m(z) \geq \beta(z) \geq \lambda_1 \text{ a.e. on } Z$$
and by hypothesis $(H_j)^4(v)$, the last inequality is strict on a set of positive measure. Hence from the monotonicity of the eigenvalue with respect to the weight function, we have
$$\widehat{\lambda}_1(m) \leq \widehat{\lambda}_1(\beta) < \widehat{\lambda}_1(\lambda_1) = 1.$$
Combining this with (5.44), we see that $y \in C_0^1(\overline{Z})$ (by the nonlinear regularity theory) cannot be an eigenfunction corresponding to $\widehat{\lambda}_1(m)$. Therefore it must change sign, a contradiction to the fact that $y \geq 0$. This proves our claim. Then because of (5.33) and the homotopy invariance of the degree map deg, we have
$$(5.45) \qquad \deg(A - N, B_R, 0) = \deg_{(S)_+}(A - hK_-, B_R, 0) \text{ for all } R > R_0.$$
Let $g \in L^\infty(Z)_+$, $g \neq 0$. Then the operators $A - hK_-$ and $A - hK_- + g$ are $(S)_+$–homotopic via the $(S)_+$–homotopy
$$h_2(t, x) = A(x) - hK_-(x) + tg \text{ for all } (t, x) \in [0, 1] \times W_0^{1,p}(Z).$$

Suppose that $h_2(t,x) = 0$ for some $t \in [0,1]$ and some $x \neq 0$. Then
$$A(x) = hK_-(x) - tg.$$

First we suppose that $t = 0$. Then we have

(5.46) $$A(x) = hK_-(x).$$

We act with the test function $x^+ \in W_0^{1,p}(Z)$ and obtain
$$\|Dx^+\|_p = 0$$
hence
$$x \leq 0.$$

Also from (5.46), we have
$$-\triangle_p x(z) = (-h(z))|x^-(z)|^{p-2} x(z) \text{ a.e. on } Z.$$

But from the chooice of h, we have
$$\widehat{\lambda}_1(-h) < \widehat{\lambda}_1(\lambda_1) = 1.$$

and so x must change sign, a contradiction to the fact that $x \leq 0$. Suppose that $t \in (0,1]$ and

(5.47) $$A(x) = hK_-(x) - tg.$$

We act with the test function $x^+ \in W_0^{1,p}(Z)$. Since $g \geq 0$, we obtain
$$\|Dx^+\|_p = 0$$
hence
$$x \leq 0.$$

From (5.47), we have
$$-\triangle_p x(z) = (-h(z))|x^-(z)|^{p-2} x(z) - tg(z) \text{ a.e. on } Z,$$
and therefore

(5.48) $$-\triangle_p(-x)(z) = (-h(z))|(-x)(z)|^{p-2}(-x)(z) + tg(z) \text{ a.e. on } Z.$$

Since $-h, g, -x \geq 0$, we infer that
$$\triangle_p(-x)(z) \leq 0 \text{ a.e. on } Z.$$

From nonlinear regularity theory, we infer that $x \in C_0^1(\overline{Z})$ (see [**17**], p. 737-738). So we can apply the strict maximum principle of Vazquez [**45**] and obtain
$$(-x) \in intC_0^1(\overline{Z})_+.$$

On the other hand, since $\widehat{\lambda}_1(-h) < 1$, Proposition 4.1 and Remark 5.5 of [**19**], imply that (5.48) cannot have a positive solution, a contradiction. Therefore we have verified that $h_2(t,x) \neq 0$ for all $t \in [0,1]$ and all $x \neq 0$. The homotopy invariance of the degree map implies that
$$\deg_{(S)_+}(A - hK_-, B_R, 0) = \deg_{(S)_+}(A - hK_- + g, B_R, 0), \forall R > 0.$$

But it is clear from the above arguments that
$$\deg_{(S)_+}(A - hK_- + g, B_R, 0) = 0$$
hence

(5.49) $$\deg_{(S)_+}(A - hK_-, B_R, 0) = 0.$$

So finally (cf. (5.45))
$$\deg(A - N, B_R, 0) = 0 \text{ for all } R \geq R_0.$$

□

Now we prove a similar result for small balls.

PROPOSITION 31. *If hypotheses* $(H_j)^4$ *hold, then there exists* $\rho_0 > 0$ *such that*
$$\deg(A - N, B_\rho, 0) = 0 \text{ for all } 0 < \rho \leq \rho_0.$$

PROOF. In this case, we consider the bounded, continuous, strictly monotone (hence maximal monotone, too) operator $K_+ : L^p(Z) \to L^q(Z)$ defined by
$$K_+(x)(.) = |x^+(.)|^{p-2} x^+(.)$$
(recall that $x^+ = \max\{x, 0\}$ and if $x \in W_0^{1,p}(Z)$, then $x^+ \in W_0^{1,p}(Z)$). This map, viewed from $W_0^{1,p}(Z)$ into $W^{-1,q}(Z) = W_0^{1,p}(Z)^*$, is completely continuous, hence compact. We consider the admissible homotopy $h_3 : [0,1] \times W_0^{1,p}(Z) \to 2^{W^{-1,q}(Z)} \setminus \{\varnothing\}$ defined by
$$h_3(t, x) = Ax - (1-t)N(x) - tg_0 K_+(x) \text{ for all } (t, x) \in [0,1] \times W_0^{1,p}(Z),$$
where $g_0 \in L^\infty(Z)_+$ with $\eta(z) \leq g_0(z) \leq \widehat{\eta}(z)$ a.e. on Z (see $(H_j)^4$ (vi)).
Claim: There exists $\rho_0 > 0$ such that for all $0 < \rho \leq \rho_0$ we have

(5.50) $\quad 0 \notin h_3(t, x)$ for all $t \in [0, 1]$ and all $x \in \partial B_\rho(0)$.

As before we suppose that the Claim is not true. Then we can find $\{x_n\}_{n \geq 1} \subseteq W_0^{1,p}(Z)$ and $\{t_n\}_{n \geq 1} \subseteq [0, 1]$ such that
$$t_n \to t \in [0, 1], \quad \|x_n\| \to 0 \text{ as } n \to \infty$$
and

(5.51) $\quad Ax_n = (1 - t_n)u_n + t_n g_0 K_+(x_n)$ with $u_n \in N(x_n)$ for all $n \geq 1$.

We set
$$y_n = \frac{x_n}{\|x_n\|}, \quad n \geq 1$$
and we may assume (at least for a subsequence) that
$$y_n \xrightarrow{w} y \text{ in } W_0^{1,p}(Z) \text{ and } y_n \to y \text{ in } L^p(Z) \text{ as } n \to \infty.$$
We divide (5.51) with $\|x_n\|^{p-1}$ and we obtain
$$A(y_n) = (1 - t_n)\frac{u_n}{\|x_n\|^{p-1}} + t_n g_0 K_+(y_n).$$

Because of hypothesis $(H_j)^4$ (vi), we can find $\delta_1 > 0$ such that

(5.52) $\quad -1 \leq \dfrac{u}{|x|^{p-2} x} \leq \widehat{\eta}(z) + 1$ for a.a. $z \in Z$, all $|x| \leq \delta_1$ and all $u \in \partial j(z, x)$.

On the other hand,, hypothesis $(H_j)^4$ (iii) implies that
$$|u| \leq a(z) + c|x|^{p-1} \text{ for a.a. } z \in Z, \text{ all } |x| > \delta_1 \text{ and all } u \in \partial j(z, x),$$

hence

(5.53) $|u| \leq \left(\dfrac{a(z)}{\delta_1^{p-1}} + c\right) |x|^{p-1}$ for a.a. $z \in Z$, all $|x| > \delta_1$ and all $u \in \partial j(z, x)$.

From (5.52) and (5.53), it follows that
(5.54)
$$|u| \leq c_1 |x|^{p-1} \text{ for some } c_1 > 0, \text{ a.a. } z \in Z, \text{ all } x \in \mathbb{R} \text{ and all } u \in \partial j(z, x).$$

Using (5.54), we have
$$\int_Z \dfrac{|u_n(z)|^q}{\|x_n\|^p} dz \leq c_1 \|y_n\|_p^p \text{ for all } n \geq 1,$$

hence
$$\left\{\dfrac{u_n}{\|x_n\|^{p-1}}\right\}_{n \geq 1} \subseteq L^q(Z)$$

is bounded. So we may assume that
$$\dfrac{u_n}{\|x_n\|^{p-1}} \xrightarrow{w} f \text{ in } L^q(Z) \text{ as } n \to \infty.$$

Given $\varepsilon > 0$ and $n \geq 1$, we consider the sets
$$C^+_{\varepsilon,n} = \left\{z \in Z : x_n > 0, \ \eta(z) - \varepsilon \leq \dfrac{u_n(z)}{x_n(z)^{p-1}} \leq \widehat{\eta}(z) + \varepsilon\right\}$$

and
$$C^-_{\varepsilon,n} = \left\{z \in Z : x_n < 0, \ -\varepsilon \leq \dfrac{u_n(z)}{|x_n(z)|^{p-2} x_n(z)} \leq \varepsilon\right\}.$$

Since $\|x_n\| \to 0$, by passing to a subsequence if necessary, we may also assume that
$$x_n(z) \to 0 \text{ a.e. on } Z.$$

Then using the sets $C^+_{\varepsilon,n}$ and arguing as in the proof of Proposition 14, we obtain
(5.55) $\eta(z) y(z)^{p-1} \leq f(z) \leq \widehat{\eta}(z) y(z)^{p-1}$ a.e. on $\{y > 0\}$.

Similarly, using the sets $C^-_{\varepsilon,n}$, we obtain
(5.56) $f(z) = 0$ a.e. on $\{y < 0\}$.

Finally from (5.54) it is clear that
(5.57) $f(z) = 0$ a.e. on $\{y = 0\}$.

Combining (5.55), (5.56) and (5.57), we conclude that
(5.58) $\eta(z) y^+(z)^{p-1} \leq f(z) \leq \widehat{\eta}(z) y^+(z)^{p-1}$ a.e. on Z,

hence
(5.59) $f(z) = g(z) y^+(z)^{p-1}$ with $g \in L^\infty(Z)_+$ such that.
$\eta(z) \leq g(z) \leq \widehat{\eta}(z)$ a.e. on Z.

Recall that
(5.60) $A(y_n) = (1 - t_n) \dfrac{u_n}{\|x_n\|^{p-1}} + t_n g_0 K_+(y_n)$ for all $n \geq 1$,

hence
$$\langle A(y_n), y_n - y \rangle = (1 - t_n) \int_Z \frac{u_n}{\|x_n\|^{p-1}} (y_n - y) \, dz + t_n \int_Z g_0 (y_n^+)^{p-1} (y_n - y) \, dz.$$

Evidently, we have
$$(1 - t_n) \int_Z \frac{u_n}{\|x_n\|^{p-1}} (y_n - y) \, dz \to 0$$

and
$$t_n \int_Z g_0 (y_n^+)^{p-1} (y_n - y) \, dz \to 0$$

as $n \to \infty$. Therefore, it follows that
$$\lim_{n \to \infty} \langle A(y_n), y_n - y \rangle = 0.$$

Since A is of type $(S)_+$ (see Lemma 11), it follows that $y_n \to y$ in $W_0^{1,p}(Z)$, hence

(5.61) $\qquad \|y\| = 1$, i.e., $y \neq 0$.

Passing to the limit as $n \to \infty$ in (5.60), we obtain
$$A(y) = (1-t) g (y^+)^{p-1} + t_n g_0 K_+(y),$$

hence
$$A(y) = \widehat{g} (y^+)^{p-1} \text{ with } \widehat{g} = (1-t) g + t g_0 \in L^\infty(Z)_+.$$

Acting with the test function $-y^- \in W_0^{1,p}(Z)$, we obtain
$$\|Dy^-\|_p = 0.$$

Recall that
$$Dy^-(z) = \begin{cases} -Dy(z) & \text{if } y(z) < 0 \\ 0 & \text{if } y(z) \geq 0. \end{cases}$$

So $y^- = 0$, hence $y \geq 0$. Therefore
$$A(y) = \widehat{g} y^{p-1},$$

hence

(5.62) $\qquad \begin{cases} -\operatorname{div}\left(\|Dy(z)\|_{\mathbb{R}^N}^{p-2} Dy(z)\right) = \widehat{g}(z) |y(z)|^{p-2} y(z) \text{ a.e. on } Z, \\ y \equiv 0 \text{ on } \partial Z. \end{cases}$

From the definition of $\widehat{g}$ we have
$$\eta(z) \leq \widehat{g}(z) \leq \widehat{\eta}(z) \text{ a.e. on } Z,$$

hence

(5.63) $\qquad \widehat{\lambda}_1(\widehat{g}) \leq \widehat{\lambda}_1(\eta) < \widehat{\lambda}_1(\lambda_1) = 1.$

Because of (5.62) and (5.63) and since $y \neq 0$, we infer that $y \in C_0^1(\overline{Z})$ (cf. the nonlinear regularity theory) must change sign, which contradicts the fact that $y \geq 0$. This proves the Claim. Then, because of (5.50) and the homotopy invariance of the degree map deg, we have

(5.64) $\qquad \deg(A - N, B_\rho, 0) = \deg_{(S)_+}(A - g_0 K_+, B_\rho, 0)$ for all $0 < \rho < \rho_0$.

Now consider $h \in L^\infty(Z)_+ \setminus \{0\}$. Then the $(S)_+$ operators $A - g_0 K_+$ and $A - g_0 K_+ - h$ are $(S)_+$-homotopic via the $(S)_+$-homotopy
$$h_4(t, x) = A(x) - g_0 K_+(x) - th \text{ for all } (t, x) \in [0, 1] \times W_0^{1,p}(Z).$$

5. HEMIVARIATIONAL INEQUALITIES WITH AN ASYMMETRIC SUBDIFFERENTIAL

As in the proof of Proposition 30 we can check that $h_4(t,x) \neq 0$ for all $t \in [0,1]$ and all $x \neq 0$. So by virtue of the homotopy invariance of the degree map $\deg_{(S)_+}$, we have

(5.65) $$\deg_{(S)_+} (A - g_0 K_+, B_\rho, 0) = \deg_{(S)_+} (A - g_0 K_+ - h, B_\rho, 0).$$

Suppose that
$$\deg_{(S)_+} (A - g_0 K_+ - h, B_\rho, 0) \neq 0.$$

Then we can find $x \in \overline{B}_\rho \setminus \{0\}$ such that

(5.66) $$A(x) = g_0 K_+ + h.$$

Using as a test function $-x^- \in W_0^{1,p}(Z)$, we obtain

(5.67) $$\|Dx^-\|_p = 0, \text{ i.e., } x \geq 0.$$

Also from (5.66) we have
(5.68)
$$\begin{cases} -\operatorname{div}\left(\|Dx(z)\|_{\mathbb{R}^N}^{p-2} Dx(z)\right) = g_0(z) |x(z)|^{p-2} x(z) + h(z) \text{ a.e. on } Z, \\ x \equiv 0 \text{ on } \partial Z. \end{cases}$$

Taking into account (5.67), (5.68) and the fact that $\widehat{\lambda}_1(g_0) < 1$, we contradict Proposition 4.1 of [19] (see also [19], Remark 5.5). Therefore, we must have
$$\deg_{(S)_+} (A - g_0 K_+ - h, B_\rho, 0) = 0,$$
hence
$$\deg(A - N, B_\rho, 0) = 0 \text{ for all } 0 \leq \rho \leq \rho_0$$
(see (5.64) and (5.65)). $\square$

Now, we are in a position to prove the multiplicity result for problem (5.1).

THEOREM 32. *If hypotheses* $(H_j)^4$ *hold, then problem* (5.1) *has at least two nontrivial solutions* $x_0, \widehat{x} \in C_0^1(\overline{Z})$ *with* $x_0 \in \operatorname{int} C_0^1(Z)_+$.

PROOF. One solution $x_0 \in \operatorname{int} C_0^1(Z)_+$ was obtained in Proposition 28. Next, let $r > 0$ be as in Proposition 29. Then choose $R > 0$ large (as in Proposition 30), and $\rho > 0$ small (as in Proposition 31) such that $B_\rho(0) \cap B_r(x_0) = \emptyset$ and $B_r(x_0) \subseteq B_R(0)$. From the additivity property of the degree map deg and Propositions 28, 29 and 30, we have

$$0 = \deg(A - N, B_R, 0) = \deg(A - N, B_r(x_0), 0) + \deg(A - N, B_\rho(0), 0)$$
$$+ \deg\left(A - N, \overline{B_R(0) \setminus B_r(x_0) \cup B_\rho(0)}, 0\right)$$
$$= 1 + \deg\left(A - N, \overline{B_R(0) \setminus B_r(x_0) \cup B_\rho(0)}, 0\right),$$

hence
$$\deg\left(A - N, \overline{B_R(0) \setminus B_r(x_0) \cup B_\rho(0)}, 0\right) = -1.$$

So, invoking the solution property of the degree map deg, we infer that there exists
$$\widehat{x} \in \overline{B_R(0) \setminus B_r(x_0) \cup B_\rho(0)},$$
hence $\widehat{x} \neq x_0$, $\widehat{x} \neq 0$, such that
$$A(\widehat{x}) = \widehat{u} \text{ with } \widehat{u} \in N(\widehat{x}).$$

From this operator equation, it follows $\widehat{x} \in W_0^{1,p}(Z)$ is a nontrivial solution of problem(5.1) and then, with the help of the nonlinear regularity theory (see Gasinski-Papageorgiou [**18**], pp. 115-116) we conclude that $\widehat{x} \in C_0^1(\overline{Z})$. □

Bibliography

[1] H. Amann. A note on degree theory for gradient mappings. *Proc. Amer. Math. Soc.*, 85:591–595, 1982.

[2] H. Amann and E. Zehnder. Nontrivial solutions for a class of nonresonance problems and applications to nonlinear differential equations. *Ann. Sc. Norm. Super. Pisa Cl. Sci.*, 7:539–603, 1980.

[3] A. Anane. Simplicité et isolation de la première valeur propre du p-Laplacian avec poids. *C. R. Math. Acad. Sci. Paris*, 305:725–728, 1987.

[4] A. Anane and N. Tsouli. On the second eigenvalue of the p-Laplacian. In *Nonlinear Partial Differential Equations*, volume 343 of *Pitman Res. Notes Math. Series*, pages 1–9. Longman, Harlow, 1996.

[5] V. Barbu. *Nonlinear Semigroups and Differential Equations in Banach Spaces*. Noordhoff International Publ., Leyden, 1976.

[6] T. Bartsch and S. Li. Critical point theory for asymptotically quadratic functionals and applications to problems with resonance. *Nonlinear Anal.*, 28:419–441, 1997.

[7] H. Brézis and L. Nirenberg. H^1 versus C^1 local minimizers. *C. R. Math. Acad. Sci. Paris*, t. 317:465–472, 1993.

[8] F. Browder. Fixed point theory and nonlinear problems. *Bull. Amer. Math. Soc.*, 9:1–39, 1983.

[9] A. Cellina. Approximation of set-valued functions and fixed point theorems. *Ann. Mat. Pura Appl.*, 82:17–24, 1969.

[10] F. H. Clarke. *Optimization and Nonsmooth Analysis*. Wiley, New York, 1983.

[11] E. Dancer and Z. Zhang. Fucik spectrum, sign changing and multiple solutions for semilinear elliptic boundary value problems with resonance at infinity. *J. Math. Anal. Appl.*, 250:449–464, 2000.

[12] Z. Denkowski, S. Migorski, and N. S. Papageorgiou. *An Introduction to Nonlinear Analysis: Applications*. Kluwer/Plenum, New York, 2003.

[13] Z. Denkowski, S. Migorski, and N. S. Papageorgiou. *An Introduction to Nonlinear Analysis: Theory*. Kluwer/Plenum, New York, 2003.

[14] G. Dinca, P. Jebelean, and J. Mawhin. Variational and topological methods for Dirichlet problems with p-Laplacian. *Port. Math.*, 58:339–378, 2001.

[15] P. Drabek, A. Kufner, and F. Nicolosi. *Quasilinear Elliptic Equations with Degenerations and Singularities*. Walter de Gruyter, Berlin, 1997.

[16] X. L. Fan, Y. Z. Zhao, and G. F. Huang. Existence of solutions for the p-Laplacian with crossing nonlinearity. *Discrete Contin. Dyn. Syst.*, 8:1019–1024, 2002.

[17] L. Gasinski and N. S. Papageorgiou. *Nonlinear Analysis*. Chapman Hall and CRC Press, Boca Raton, 2005.

[18] L. Gasinski and N. S. Papageorgiou. *Nonsmooth Critical Point Theory and Nonlinear Boundary Value Problems*. Chapman Hall and CRC Press, Boca Raton, 2005.

[19] T. Godoy, J. P. Gossez, and S. Paczka. On the antimaximum principle for the p-Laplacian with indefinite weight. *Nonlinear Anal.*, 51:449–467, 2002.

[20] J. V. A. Gonçalves and O. Miyagaki. Three solutions for a strongly resonant elliptic problem. *Nonlinear Anal.*, 24:265–272, 1995.

[21] N. Hirano. Multiple solutions for quasilinear elliptic equations. *Nonlinear Anal.*, 15:625–638, 1990.

[22] N. Hirano. Existence of multiple solutions for quasilinear and semilinear elliptic equations. *Nonlinear Anal.*, 19:123–143, 1992.

[23] N. Hirano and T. Nishimura. Multiplicity results for semilinear elliptic problems at resonance with jumping nonlinearities. *J. Math. Anal. Appl.*, 180:566–586, 1993.

[24] S. Hu and N. S. Papageorgiou. Generalizations of Browder's degree theory. *Trans. Amer. Math. Soc.*, 347:233–259, 1995.

[25] S. Hu and N. S. Papageorgiou. *Handbook of Multivalued Analysis, Vol. I: Theory.* Kluwer, Dordrecht, 1997.

[26] Y. Huang and H. S. Zhou. Positive solutions for $-\triangle_p u = f(x,u)$ growing as u^{p-1} at infinity. *Appl. Math. Lett.*, 17:881–887, 2004.

[27] S. Kobayashi and M. Otani. Topological degree for $(S)_+$-mappings with maximal monotone perturbations and its applications to variational inequalities. *Nonlinear Anal.*, 59:147–172, 2004.

[28] M. Kucera. A global continuation theorem for obtaining eigenvalues and bifurcation points. *Czechoslovak Math: J.*, 38:120–137, 1988.

[29] S. Kyritsi and N. S. Papageorgiou. Hemivariational inequalities with the potential crossing te first eigenvalue. *Bull. Austral. Math. Soc.*, 64:381–393, 2001.

[30] S. Kyritsi and N. S. Papageorgiou. Multiple solutions of constant sign for nonlinear nonsmooth eigenvalue problems near resonance. *Calc. Var. Partial Differential Equations*, 20:1–24, 2004.

[31] E. Landesman, S. Robinson, and A. Rumbos. Multiple solutions of semilinear elliptic problems at resonance. *Nonlinear Anal.*, 24:1049–1059, 1995.

[32] V. K. Le. Some global bifurcation results for variational inequalities. *J. Differential Equations*, 131:39–78, 1996.

[33] V. K. Le. Some degree calculations and applications to global bifurcations of variational inequalities. *Nonlinear Anal.*, 37:473–500, 1999.

[34] G. Li and H. S. Zhou. Asymptotically linear Dirichlet problems for the p-Laplacian. *Nonlinear Anal.*, 43:1043–1055, 2001.

[35] G. Li and H. S. Zhou. Multiple solutions to p-Laplacian problems with asymptotic nonlinearity as u^{p-1} at infinity. *J. London Math. Soc.*, 65:123–138, 2002.

[36] J. Q. Liu and J. B. Su. Remarks on multiple nontrivial solutions for quasilinear resonant problems. *J. Math. Anal.Appl.*, 258:209–222, 2001.

[37] E. Miersemann. Eigenvalue problem for variational inequalities. *Contemp. Math.*, 4:25–43, Amer. Math.Soc., Providence, R. I., 1981.

[38] N. Mizoguchi. Asymptotically linear elliptic equations without nonresonance conditions. *J. Differential Equations*, 113:150–165, 1994.

[39] U. Mosco. On the continuity of the Young-Fenchel transform. *J. Math. Anal. Appl.*, 35:518–535, 1971.

[40] D. Motreanu and N. S. Papageorgiou. Multiple solutions for nonlinear elliptic equations at resonance with a nonsmooth potential. *Nonlinear Anal.*, 56:1211–1234, 2004.

[41] Z. Naniewicz and P. Panagiotopoulos. *Mathematical Theory of Hemi-Variational Inequalities and Applications.* Marcel Dekker, New York, 1995.

[42] C. Stuart and H. S. Zhou. Applying the mountain pass theorem to an asymptotically linear elliptic equation in $\mathbb{R}^n$. *Comm. Partial Differential Equations*, 24:1731–1758, 1999.

[43] A. Szulkin. Positive solutions of variational inequalities: A degree-theoretic approach. *J. Differential Equations*, 57:90–111, 1985.

[44] A. Szulkin. Minimax principles for lower semicontinuous functions and applications to nonlinear boundary value problems. *Ann. Inst. H. Poincare Anal. Non Linéaire*, 3:77–109, 1986.

[45] J. Vazquez. A strong maximum principle for some quasilinear elliptic equations. *Appl. Math. Optim.*, 12:191–202, 1984.

[46] E. Zeidler. *Nonlinear Functional Analysis and its Applications II.* Springer-Verlag, New York, 1990.

Editorial Information

To be published in the *Memoirs*, a paper must be correct, new, nontrivial, and significant. Further, it must be well written and of interest to a substantial number of mathematicians. Piecemeal results, such as an inconclusive step toward an unproved major theorem or a minor variation on a known result, are in general not acceptable for publication.

Papers appearing in *Memoirs* are generally at least 80 and not more than 200 published pages in length. Papers less than 80 or more than 200 published pages require the approval of the Managing Editor of the Transactions/Memoirs Editorial Board.

As of July 31, 2008, the backlog for this journal was approximately 16 volumes. This estimate is the result of dividing the number of manuscripts for this journal in the Providence office that have not yet gone to the printer on the above date by the average number of monographs per volume over the previous twelve months, reduced by the number of volumes published in four months (the time necessary for preparing a volume for the printer). (There are 6 volumes per year, each usually containing at least 4 numbers.)

A Consent to Publish and Copyright Agreement is required before a paper will be published in the *Memoirs*. After a paper is accepted for publication, the Providence office will send a Consent to Publish and Copyright Agreement to all authors of the paper. By submitting a paper to the *Memoirs*, authors certify that the results have not been submitted to nor are they under consideration for publication by another journal, conference proceedings, or similar publication.

Information for Authors

Memoirs are printed from camera copy fully prepared by the author. This means that the finished book will look exactly like the copy submitted.

Initial submission. The AMS uses Centralized Manuscript Processing for initial submissions. Authors should submit a PDF file using the Initial Manuscript Submission form found at www.ams.org/peer-review-submission, or send one copy of the manuscript to the following address: Centralized Manuscript Processing, MEMOIRS OF THE AMS, 201 Charles Street, Providence, RI 02904-2294 USA. If a paper copy is being forwarded to the AMS, indicate that it is for it Memoirs and include the name of the corresponding author, contact information such as email address or mailing address, and the name of an appropriate Editor to review the paper (see the list of Editors below).

The paper must contain a *descriptive title* and an *abstract* that summarizes the article in language suitable for workers in the general field (algebra, analysis, etc.). The *descriptive title* should be short, but informative; useless or vague phrases such as "some remarks about" or "concerning" should be avoided. The *abstract* should be at least one complete sentence, and at most 300 words. Included with the footnotes to the paper should be the 2000 *Mathematics Subject Classification* representing the primary and secondary subjects of the article. The classifications are accessible from www.ams.org/msc/. The list of classifications is also available in print starting with the 1999 annual index of *Mathematical Reviews*. The Mathematics Subject Classification footnote may be followed by a list of *key words and phrases* describing the subject matter of the article and taken from it. Journal abbreviations used in bibliographies are listed in the latest *Mathematical Reviews* annual index. The series abbreviations are also accessible from www.ams.org/msnhtml/serials.pdf. To help in preparing and verifying references, the AMS offers MR Lookup, a Reference Tool for Linking, at www.ams.org/mrlookup/.

Electronically prepared manuscripts. The AMS encourages electronically prepared manuscripts, with a strong preference for $\mathcal{AMS}$-LaTeX. To this end, the Society has prepared $\mathcal{AMS}$-LaTeX author packages for each AMS publication. Author packages include instructions for preparing electronic manuscripts, samples, and a style file that generates

Editorial Information

To be published in the *Memoirs*, a paper must be correct, new, nontrivial, and significant. Further, it must be well written and of interest to a substantial number of mathematicians. Piecemeal results, such as an inconclusive step toward an unproved major theorem or a minor variation on a known result, are in general not acceptable for publication.

Papers appearing in *Memoirs* are generally at least 80 and not more than 200 published pages in length. Papers less than 80 or more than 200 published pages require the approval of the Managing Editor of the Transactions/Memoirs Editorial Board.

As of July 31, 2008, the backlog for this journal was approximately 16 volumes. This estimate is the result of dividing the number of manuscripts for this journal in the Providence office that have not yet gone to the printer on the above date by the average number of monographs per volume over the previous twelve months, reduced by the number of volumes published in four months (the time necessary for preparing a volume for the printer). (There are 6 volumes per year, each usually containing at least 4 numbers.)

A Consent to Publish and Copyright Agreement is required before a paper will be published in the *Memoirs*. After a paper is accepted for publication, the Providence office will send a Consent to Publish and Copyright Agreement to all authors of the paper. By submitting a paper to the *Memoirs*, authors certify that the results have not been submitted to nor are they under consideration for publication by another journal, conference proceedings, or similar publication.

Information for Authors

Memoirs are printed from camera copy fully prepared by the author. This means that the finished book will look exactly like the copy submitted.

Initial submission. The AMS uses Centralized Manuscript Processing for initial submissions. Authors should submit a PDF file using the Initial Manuscript Submission form found at www.ams.org/peer-review-submission, or send one copy of the manuscript to the following address: Centralized Manuscript Processing, MEMOIRS OF THE AMS, 201 Charles Street, Providence, RI 02904-2294 USA. If a paper copy is being forwarded to the AMS, indicate that it is for it Memoirs and include the name of the corresponding author, contact information such as email address or mailing address, and the name of an appropriate Editor to review the paper (see the list of Editors below).

The paper must contain a *descriptive title* and an *abstract* that summarizes the article in language suitable for workers in the general field (algebra, analysis, etc.). The *descriptive title* should be short, but informative; useless or vague phrases such as "some remarks about" or "concerning" should be avoided. The *abstract* should be at least one complete sentence, and at most 300 words. Included with the footnotes to the paper should be the 2000 *Mathematics Subject Classification* representing the primary and secondary subjects of the article. The classifications are accessible from www.ams.org/msc/. The list of classifications is also available in print starting with the 1999 annual index of *Mathematical Reviews*. The Mathematics Subject Classification footnote may be followed by a list of *key words and phrases* describing the subject matter of the article and taken from it. Journal abbreviations used in bibliographies are listed in the latest *Mathematical Reviews* annual index. The series abbreviations are also accessible from www.ams.org/msnhtml/serials.pdf. To help in preparing and verifying references, the AMS offers MR Lookup, a Reference Tool for Linking, at www.ams.org/mrlookup/.

Electronically prepared manuscripts. The AMS encourages electronically prepared manuscripts, with a strong preference for $\mathcal{AMS}$-LaTeX. To this end, the Society has prepared $\mathcal{AMS}$-LaTeX author packages for each AMS publication. Author packages include instructions for preparing electronic manuscripts, samples, and a style file that generates the particular design specifications of that publication series. Though $\mathcal{AMS}$-LaTeX is the highly preferred format of TeX, author packages are also available in $\mathcal{AMS}$-TeX.

Authors may retrieve an author package for *Memoirs of the AMS* from www.ams.org/journals/memo/memoauthorpac.html or via FTP to ftp.ams.org (login as anonymous, enter username as password, and type cd pub/author-info). The *AMS Author Handbook* and the *Instruction Manual* are available in PDF format from the author package link. The author package can also be obtained free of charge by sending email to tech-support@ams.org (Internet) or from the Publication Division, American Mathematical Society, 201 Charles St., Providence, RI 02904-2294, USA. When requesting an author package, please specify $\mathcal{AMS}$-LaTeX or $\mathcal{AMS}$-TeX and the publication in which your paper will appear. Please be sure to include your complete mailing address.

After acceptance. The final version of the electronic file should be sent to the Providence office (this includes any TeX source file, any graphics files, and the DVI or PostScript file) immediately after the paper has been accepted for publication.

Before sending the source file, be sure you have proofread your paper carefully. The files you send must be the EXACT files used to generate the proof copy that was accepted for publication. For all publications, authors are required to send a printed copy of their paper, which exactly matches the copy approved for publication, along with any graphics that will appear in the paper.

Accepted electronically prepared files can be submitted via the web at www.ams.org/submit-book-journal/, sent via FTP, or sent on CD-Rom or diskette to the Electronic Prepress Department, American Mathematical Society, 201 Charles Street, Providence, RI 02904-2294 USA. TeX source files, DVI files, and PostScript files can be transferred over the Internet by FTP to the Internet node ftp.ams.org (130.44.1.100). When sending a manuscript electronically via CD-Rom or diskette, please be sure to include a message identifying the paper as a Memoir.

Electronically prepared manuscripts can also be sent via email to pub-submit@ams.org (Internet). In order to send files via email, they must be encoded properly. (DVI files are binary and PostScript files tend to be very large.)

Electronic graphics. Comprehensive instructions on preparing graphics are available at www.ams.org/authors/journals.html. A few of the major requirements are given here.

Submit files for graphics as EPS (Encapsulated PostScript) files. This includes graphics originated via a graphics application as well as scanned photographs or other computer-generated images. If this is not possible, TIFF files are acceptable as long as they can be opened in Adobe Photoshop or Illustrator. No matter what method was used to produce the graphic, it is necessary to provide a paper copy to the AMS.

Authors using graphics packages for the creation of electronic art should also avoid the use of any lines thinner than 0.5 points in width. Many graphics packages allow the user to specify a "hairline" for a very thin line. Hairlines often look acceptable when proofed on a typical laser printer. However, when produced on a high-resolution laser imagesetter, hairlines become nearly invisible and will be lost entirely in the final printing process.

Screens should be set to values between 15% and 85%. Screens which fall outside of this range are too light or too dark to print correctly. Variations of screens within a graphic should be no less than 10%.

Inquiries. Any inquiries concerning a paper that has been accepted for publication should be sent to memo-query@ams.org or directly to the Electronic Prepress Department, American Mathematical Society, 201 Charles St., Providence, RI 02904-2294 USA.

Editors

This journal is designed particularly for long research papers, normally at least 80 pages in length, and groups of cognate papers in pure and applied mathematics. Papers intended for publication in the *Memoirs* should be addressed to one of the following editors. The AMS uses Centralized Manuscript Processing for initial submissions to AMS journals. Authors should follow instructions listed on the Initial Submission page found at www.ams.org/memo/memosubmit.html.

Algebra to ALEXANDER KLESHCHEV, Department of Mathematics, University of Oregon, Eugene, OR 97403-1222; email: ams@noether.uoregon.edu

Algebraic geometry and its application to MINA TEICHER, Emmy Noether Research Institute for Mathematics, Bar-Ilan University, Ramat-Gan 52900, Israel; email: teicher@macs.biu.ac.il

Algebraic geometry to DAN ABRAMOVICH, Department of Mathematics, Brown University, Box 1917, Providence, RI 02912; email: amsedit@math.brown.edu

Algebraic topology to ALEJANDRO ADEM, Department of Mathematics, University of British Columbia, Room 121, 1984 Mathematics Road, Vancouver, British Columbia, Canada V6T 1Z2; email: adem@math.ubc.ca

Combinatorics to JOHN R. STEMBRIDGE, Department of Mathematics, University of Michigan, Ann Arbor, Michigan 48109-1109; email: FRS@umich.edu

Complex analysis and harmonic analysis to ALEXANDER NAGEL, Department of Mathematics, University of Wisconsin, 480 Lincoln Drive, Madison, WI 53706-1313; email: nagel@math.wisc.edu

Differential geometry and global analysis to LISA C. JEFFREY, Department of Mathematics, University of Toronto, 100 St. George St., Toronto, ON Canada M5S 3G3; email: jeffrey@math.toronto.edu

Dynamical systems and ergodic theory and complex anaysis to YUNPING JIANG, Department of Mathematics, CUNY Queens College and Graduate Center, 65-30 Kissena Blvd., Flushing, NY 11367; email: Yunping.Jiang@qc.cuny.edu

Functional analysis and operator algebras to DIMITRI SHLYAKHTENKO, Department of Mathematics, University of California, Los Angeles, CA 90095; email: shlyakht@math.ucla.edu

Geometric analysis to WILLIAM P. MINICOZZI II, Department of Mathematics, Johns Hopkins University, 3400 N. Charles St., Baltimore, MD 21218; email: trans@math.jhu.edu

Geometric analysis to MARK FEIGHN, Math Department, Rutgers University, Newark, NJ 07102; email: feighn@andromeda.rutgers.edu

Harmonic analysis, representation theory, and Lie theory to ROBERT J. STANTON, Department of Mathematics, The Ohio State University, 231 West 18th Avenue, Columbus, OH 43210-1174; email: stanton@math.ohio-state.edu

Logic to STEFFEN LEMPP, Department of Mathematics, University of Wisconsin, 480 Lincoln Drive, Madison, Wisconsin 53706-1388; email: lempp@math.wisc.edu

Number theory to JONATHAN ROGAWSKI, Department of Mathematics, University of California, Los Angeles, CA 90095; email: jonr@math.ucla.edu

Partial differential equations to GUSTAVO PONCE, Department of Mathematics, South Hall, Room 6607, University of California, Santa Barbara, CA 93106; email: ponce@math.ucsb.edu

Partial differential equations and dynamical systems to PETER POLACIK, School of Mathematics, University of Minnesota, Minneapolis, MN 55455; email: polacik@math.umn.edu

Probability and statistics to RICHARD BASS, Department of Mathematics, University of Connecticut, Storrs, CT 06269-3009; email: bass@math.uconn.edu

Real analysis and partial differential equations to DANIEL TATARU, Department of Mathematics, University of California, Berkeley, Berkeley, CA 94720; email: tataru@math.berkeley.edu

All other communications to the editors should be addressed to the Managing Editor, ROBERT GURALNICK, Department of Mathematics, University of Southern California, Los Angeles, CA 90089-1113; email: guralnic@math.usc.edu.

Titles in This Series

918 **Jonathan Brundan and Alexander Kleshchev,** Representations of shifted Yangians and finite W-algebras, 2008

917 **Salah-Eldin A. Mohammed, Tusheng Zhang, and Huaizhong Zhao,** The stable manifold theorem for semilinear stochastic evolution equations and stochastic partial differential equations, 2008

916 **Yoshikata Kida,** The mapping class group from the viewpoint of measure equivalence theory, 2008

915 **Sergiu Aizicovici, Nikolaos S. Papageorgiou, and Vasile Staicu,** Degree theory for operators of monotone type and nonlinear elliptic equations with inequality constraints, 2008

914 **E. Shargorodsky and J. F. Toland,** Bernoulli free-boundary problems, 2008

913 **Ethan Akin, Joseph Auslander, and Eli Glasner,** The topological dynamics of Ellis actions, 2008

912 **Igor Chueshov and Irena Lasiecka,** Long-time behavior of second order evolution equations with nonlinear damping, 2008

911 **John Locker,** Eigenvalues and completeness for regular and simply irregular two-point differential operators, 2008

910 **Joel Friedman,** A proof of Alon's second eigenvalue conjecture and related problems, 2008

909 **Cameron McA. Gordon and Ying-Qing Wu,** Toroidal Dehn fillings on hyperbolic 3-manifolds, 2008

908 **J.-L. Waldspurger,** L'endoscopie tordue n'est pas si tordue, 2008

907 **Yuanhua Wang and Fei Xu,** Spinor genera in characteristic 2, 2008

906 **Raphaël S. Ponge,** Heisenberg calculus and spectral theory of hypoelliptic operators on Heisenberg manifolds, 2008

905 **Dominic Verity,** Complicial sets characterising the simplicial nerves of strict ω-categories, 2008

904 **William M. Goldman and Eugene Z. Xia,** Rank one Higgs bundles and representations of fundamental groups of Riemann surfaces, 2008

903 **Gail Letzter,** Invariant differential operators for quantum symmetric spaces, 2008

902 **Bertrand Toën and Gabriele Vezzosi,** Homotopical algebraic geometry II: Geometric stacks and applications, 2008

901 **Ron Donagi and Tony Pantev (with an appendix by Dmitry Arinkin),** Torus fibrations, gerbes, and duality, 2008

900 **Wolfgang Bertram,** Differential geometry, Lie groups and symmetric spaces over general base fields and rings, 2008

899 **Piotr Hajłasz, Tadeusz Iwaniec, Jan Malý, and Jani Onninen,** Weakly differentiable mappings between manifolds, 2008

898 **John Rognes,** Galois extensions of structured ring spectra/Stably dualizable groups, 2008

897 **Michael I. Ganzburg,** Limit theorems of polynomial approximation with exponential weights, 2008

896 **Michael Kapovich, Bernhard Leeb, and John J. Millson,** The generalized triangle inequalities in symmetric spaces and buildings with applications to algebra, 2008

895 **Steffen Roch,** Finite sections of band-dominated operators, 2008

894 **Martin Dindoš,** Hardy spaces and potential theory on C^1 domains in Riemannian manifolds, 2008

893 **Tadeusz Iwaniec and Gaven Martin,** The Beltrami Equation, 2008

892 **Jim Agler, John Harland, and Benjamin J. Raphael,** Classical function theory, operator dilation theory, and machine computation on multiply-connected domains, 2008

TITLES IN THIS SERIES

891 **John H. Hubbard and Peter Papadopol,** Newton's method applied to two quadratic equations in $\mathbb{C}^2$ viewed as a global dynamical system, 2008

890 **Steven Dale Cutkosky,** Toroidalization of dominant morphisms of 3-folds, 2007

889 **Michael Sever,** Distribution solutions of nonlinear systems of conservation laws, 2007

888 **Roger Chalkley,** Basic global relative invariants for nonlinear differential equations, 2007

887 **Charlotte Wahl,** Noncommutative Maslov index and eta-forms, 2007

886 **Robert M. Guralnick and John Shareshian,** Symmetric and alternating groups as monodromy groups of Riemann surfaces I: Generic covers and covers with many branch points, 2007

885 **Jae Choon Cha,** The structure of the rational concordance group of knots, 2007

884 **Dan Haran, Moshe Jarden, and Florian Pop,** Projective group structures as absolute Galois structures with block approximation, 2007

883 **Apostolos Beligiannis and Idun Reiten,** Homological and homotopical aspects of torsion theories, 2007

882 **Lars Inge Hedberg and Yuri Netrusov,** An axiomatic approach to function spaces, spectral synthesis and Luzin approximation, 2007

881 **Tao Mei,** Operator valued Hardy spaces, 2007

880 **Bruce C. Berndt, Geumlan Choi, Youn-Seo Choi, Heekyoung Hahn, Boon Pin Yeap, Ae Ja Yee, Hamza Yesilyurt, and Jinhee Yi,** Ramanujan's forty identities for Rogers-Ramanujan functions, 2007

879 **O. García-Prada, P. B. Gothen, and V. Muñoz,** Betti numbers of the moduli space of rank 3 parabolic Higgs bundles, 2007

878 **Alessandra Celletti and Luigi Chierchia,** KAM stability and celestial mechanics, 2007

877 **María J. Carro, José A. Raposo, and Javier Soria,** Recent developments in the theory of Lorentz spaces and weighted inequalities, 2007

876 **Gabriel Debs and Jean Saint Raymond,** Borel liftings of Borel sets: Some decidable and undecidable statements, 2007

875 **C. Krattenthaler and T. Rivoal,** Hypergéométrie et fonction zêta de Riemann, 2007

874 **Sonia Natale,** Semisolvability of semisimple Hopf algebras of low dimension, 2007

873 **A. J. Duncan,** Exponential genus problems in one-relator products of groups, 2007

872 **Anthony V. Geramita, Tadahito Harima, Juan C. Migliore, and Yong Su Shin,** The Hilbert function of a level algebra, 2007

871 **Pascal Auscher,** On necessary and sufficient conditions for L^p-estimates of Riesz transforms associated to elliptic operators on $\mathbb{R}^n$ and related estimates, 2007

870 **Takuro Mochizuki,** Asymptotic behaviour of tame harmonic bundles and an application to pure twistor D-modules, Part 2, 2007

869 **Takuro Mochizuki,** Asymptotic behaviour of tame harmonic bundles and an application to pure twistor D-modules, Part 1, 2007

868 **Gelu Popescu,** Entropy and multivariable interpolation, 2006

867 **Vilmos Totik,** Metric properties of harmonic measures, 2006

866 **William Craig,** Semigroups underlying first-order logic, 2006

865 **Nathanial P. Brown,** Invariant means and finite representation theory of $C*$-algebras, 2006

864 **John M. Lee,** Fredholm operators and Einstein metrics on conformally compact manifolds, 2006

For a complete list of titles in this series, visit the
AMS Bookstore at **www.ams.org/bookstore/**.